# THE COLORADO RIVER

## Instability and Basin Management

### William L. Graf
Department of Geography
Arizona State University
Tempe

RESOURCE PUBLICATIONS
IN GEOGRAPHY

**Association of American Geographers**
**1710 Sixteenth Street N.W.**
**Washington, D.C. 20009**


Library of Congress Card Number 84-28336
ISBN 0-89291-186-7



**Library of Congress Cataloging in Publication Data**

Graf, William L, 1947-
    The Colorado River.

    (Resource publications in geography)
    Bibliography: p.
    1. Colorado River (Colo. — Mexico) — Channel.
2. Stream conservation — Colorado River (Colo.-Mexico)
3. Water resources development — Colorado River (Colo.-Mexico)
    I. Title. II. Series.
GB566.C7G7 1985      33.91'62'097913      84-28336
ISBN 0-89291-186-7



*Publication Supported by the A.A.G.*

Graphic Design by CGK

Cover photograph: Hoover Dam and Lake Mead (Bureau of Reclamation photograph
P45-300-01465). Back cover: The Colorado River Basin (redrawn from map provided by the U.S.
Bureau of Reclamation).

*Printed by Commercial Printing Inc.*
*State College, Pennsylvania*

# Foreword

During the coming decades, one of the most dramatic natural resource management dilemmas in the United States will unfold in the Southwest. Completion of the massive Central Arizona Project will finally bring actual use of the Colorado River up to the magnitude of water allocated within the basin states and to Mexico. Unfortunately, allocations were made during an unusually moist period, exceeding the long term average flow of the stream by at least 15 percent. Continued Sunbelt growth and the resulting thirst for water resources is alone a challenge to water and related land resource management in the Southwest; when promises exceed potentials for water delivery, litigation and perhaps economic and environmental crises may follow.

The Colorado Basin is characterized by instability. As Will Graf argues in *The Colorado River, Instability and Basin Management,* various forms of natural and human-induced variability are significant challenges to management strategies and control technologies. Among these are hydrological, ecological, channel, system-wide, and chemical instabilities. Analysis of these phenomena indicates the complexity of interactions among components of the river basin as a human-environmental system. Graf brings particular skills in fluvial geomorphology to his general geographical perspective on the Colorado. Thus his analysis is strengthened by technical understanding as well as geographical insights into the Colorado River basin as a whole.

The **Resource Publications in Geography** are sponsored by the Association of American Geographers, a professional organization whose purpose is to advance studies in geography and to encourage application of geographic research in education, government, and business. Tracing its origins to the AAG's *Resource Papers* (1968-1974), the **Resource Publications** continue a tradition of presenting geographers' views on timely public and professional issues to colleagues, students, and fellow citizens. Views expressed, of course, are the author's and do not imply AAG endorsement.

As issues related to the Colorado become more strident and challenging in coming years, the editor and advisory board hope that this book will provide a background to the issues as they emerge. We also hope that the book will strengthen your appreciation of a geographer's approach to river basin management.

C. Gregory Knight, *The Pennsylvania State University*
Editor, Resource Publications in Geography

## Resource Publications Advisory Board

James S. Gardner, *University of Waterloo*
Patricia Gober, *Arizona State University*
Charles M. Good, Jr., *Virginia Polytechnic Institute and State University*
Sam B. Hilliard, *Louisiana State University*
Phillip C. Muehrcke, *University of Wisconsin, Madison*
Thomas J. Wilbanks, *Oak Ridge National Laboratory*

# Preface and Acknowledgements

Several years ago in a discussion with Dick Reeves, a geographer at the University of Arizona in Tucson, I indicated that I was interested in writing a book about the Colorado River Basin. He offered some wise council: "We've got too many regional descriptions and histories — write about the principles of natural and social science that can be learned from the basin." After a number of years of developing those principles, I was afforded the opportunity to act on Reeves' suggestion by Greg Knight in the form of an invitation to write this book. I am grateful to them for their support.

The book has three general purposes. First, the volume is designed to provide university students, instructors, and professional geographers outside the academic community with an exposition of present understanding of the spatial characteristics of some environmental processes and problems in the management of an arid-region river basin. Second, the volume is intended for the informed layperson desiring information on these subjects without a laborious search through technical literature. Finally, the volume affords an opportunity to draw together recent relevant literature that has appeared in widely divergent journals in a number of disciplines. Given the intended audience of this work, mathematical and statistical aspects of the subject are not generally given, but are available in the references cited.

As a field geomorphologist interested in spatial analysis, my perspective on the management of the Colorado River Basin probably differs from that of most previous authors. This perspective has been shaped by interaction with my scientific, managerial, and legal colleagues. Like the first geomorphologists who extensively studied the basin — John Wesley Powell, Grove Karl Gilbert, and Clarence Dutton — I find that my ideas are not necessarily always my own, but rather shared with others to the point where origins become obscure. For that reason I owe a sincere thanks to my colleagues at Arizona State University: Tony Brazel, Andrew Carleton, Pat Gober, Mel Marcus, and Duncan Patten; to Stan Schumm of Colorado State University; to Vic Baker and Bill Bull of the University of Arizona; and to Ned Andrews, John Costa, Richard Hereford, and Ray Turner of the U.S. Geological Survey.

Lynn Almer, Steve Cochran, Joe Dixon, and Don Gross of the U. S. Army Corps of Engineers, Bob Fraley and Jim Upton of the U. S. Department of Justice, and a series of lawyers in private practice — Harding Cure, Doug Irish, Pamela Kingsley, Russ Piccolli, and Carlos Ronstadt — contributed greatly to my education about the application of science to problems of public policy and environmental management. Bruce Rhoads and Herbert Verville were the cartographic assistants for this volume. Debra Daggs provided valuable editorial assistance. To all of them I express my sincere appreciation.

Financial support for some of the research reported in the following pages was generously provided by the National Geographic Society (Grants 2437A and 2437B) and the National Science Foundation (Grants EAR7727698 and EAR8119932).

William L. Graf

# Contents

# List of Figures

# List of Tables

# 1

# The Colorado River Basin

In 1905, communities along the lower Colorado, Salt, and Gila rivers experienced flood damage that cost more than $3 million to repair. Portions of towns were washed away as the unstable channel migrated across floodplains, and fields were inundated by water and sediment. On the Colorado River, water broke through the headgates of the Alamo Canal causing the river to change its course and to empty almost its entire ' flow into southern California's Imperial Valley, creating the Salton Sea. In the early 1980s, after decades of analysis, planning, and construction of flood control and irrigation works costing over $1.7 billion, a crippling drought in the same region was ended by unexpected floods which caused $200 million in damage to communities along the rivers. The Colorado, a resource that provided water to nourish a multi-billion dollar agriculture system and urban developments serving more than 12 million people, had once again become an unpredictable hazard. Can we develop a geographic perspective on the interaction between an unstable, arid-region river system and a society that requires environmental stability for survival?

The clash between environmental processes and human management strategies in the Colorado River Basin provides valuable generalizations about how societies deal with adverse environments. The importance of these generalizations far outweighs the unique characteristics of some aspects of the American Southwest. In the pages that follow, three major themes emerge from the spatial perspective on the society/ environment interactions in the basin: 1) research and management strategies based on assumed predictable averages are not useful when the environment is unstable; 2) the problems created by the unstable environments are inherently spatial or geographic; and 3) a geographic perspective that includes radical variations is therefore essential to successful solutions to these problems.

## Prehistory to Science

Before the arrival of the technology-laden Anglo settlers in the Colorado River Basin in the 1800s, extensive development of water resources had already given rise to irrigation agriculture and urbanization (Figure 1). After about 1000 AD the Indian inhabitants of the area began to divert water from several rivers in the basin for growing crops of corn, beans, and squash. By 1200 AD sophisticated canal systems carried water to fields on the floodplains of the Salt, Gila, San Juan, Fremont, and other streams. Cities of up to 5,000 inhabitants sprang up in Arizona and New Mexico. Adobe houses, in some cases with a hundred or more rooms, esplanades, ball courts, and special religious structures were common features of these early urban centers. By 1400 AD the farmers and the urbanites had vanished, however, victims of an unknown combination of intense drought, floods, salinization of the fields, invasion by outsiders, or internal political dissension (for an overview, see Haury 1976; Ambler 1977).

These earliest water managers in the Colorado Basin became masters of their craft as they built irrigation-based societies in an otherwise harsh environment. Their canals were V-shaped for maximum efficiency in the movement of water with a minimum of evaporation loss and a maximum of velocity. They built storage reservoirs which lengthened the irrigation season by storing water during spring runoff maximums to be used during the later dry summer months. They built stilling basins, small holding tanks where turbulent flow could be smoothed and sediment would be deposited, thus saving the canals from bothersome siltation. They built water distribution systems so accurately surveyed that when their fields were occupied by Anglo settlers 500 years later the same alignments were used (Corle 1951; Faulk 1970).

Unlike the Anglos who came later, however, the early Indian residents viewed themselves as part of the natural environment rather than owners or mere users. Although many details of the culture of the Hohokam, Mogollon, and Anassazi peoples are not known, their art and their descendants (the Hopi of Arizona and the Pueblo of New Mexico) provide clues about their perspectives. They viewed the earth as a mother figure and themselves as children who were expected to treat their parent with respect (Waters 1963). Rather than trying radically to alter the natural environment, the early Indians attempted to accommodate their lives to the inherent variability of the world around them. They lived close to water sources and raised crops that were well-suited to the climatic, soil, and water conditions. Although they lacked the tech-

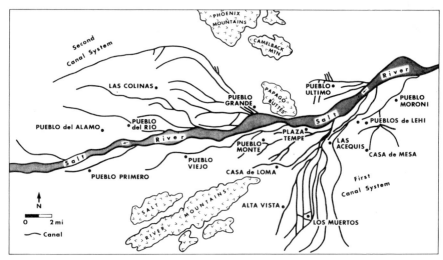

**FIGURE 1**   PREHISTORIC CANAL SYSTEMS IN THE SALT RIVER VALLEY, CENTRAL ARIZONA. Many of the canals excavated by the Hohokam Indians almost 1,000 years ago were cleared of silt and used by early Anglo settlers in the 1860s.

nology of the twentieth century, their perspective on the relationship between society and environment is one that is increasingly gaining favor among modern residents of the region.

One of the most common figures in the art of these early Indians is Kokopelle, a lone itinerant man who wandered from village to village with his ever-present flute and backpack filled with corn (Towler 1982). Kokopelle, notorious as a lover, was primarily a symbol of fertility. He saved some corn for planting rather than consuming it all, and he adopted a nomadic lifestyle suited to an unpredictable environment. His outline is still found painted or etched on canyon walls throughout the Colorado River Basin, a curiosity for a modern society that has only begun to learn the concepts of environmental accommodation which he represents.

Ancient Indians knew and understood their natural surroundings by experience and explained natural processes by their religious beliefs. Modern managers of the Colorado River Basin base their understanding on scientific methods: that is, by establishing hypotheses, collecting data, testing the hypotheses, and constructing explanatory theories. A feature common to the ancient and modern approaches is a perspective on the river basin as an holistic entity, a complex grouping of individual parts that function together. Much of our understanding of the basin processes is based on the application of general systems concepts (Chorley and Kennedy 1971).

## The Resource Base

The description of the physical characteristics of the Colorado River Basin is not the purpose of this book, but a brief review of the nature of the various elements of the basin provides a useful starting point for discussion of the system's instability (for a general map, see the back cover). The natural components of the basin that are of greatest interest are its geologic structure, geomorphology, climatology, hydrology, and ecology (Figure 2).

The Colorado River drainage area spans three significantly different geologic provinces: the Rocky Mountains, the Colorado Plateau, and the Basin and Range Province (reviewed by Hunt 1974; Coney 1983). The Rocky Mountains are masses of upthrust, highly contorted crystalline rock characterized by high altitude surfaces with thin soils. Because of their generally north-south alignment they act as a huge barrier to moisture-bearing winds from the west. Soils are poorly developed, and runoff carries relatively little sediment from this province.

The Colorado Plateau, on the other hand, is a section of uplifted and slightly deformed sedimentary rocks that form extensive mesas separated by canyons with nearly vertical walls. Volcanic features are common (Colton 1976). The sandstones and shales that dominate the surface of the plateau province shed great quantities of sediment that become a significant feature of the rivers of the basin (sediment supply reviewed by Smith *et al.* 1960; Howard and Dolan 1981). The Basin and Range Province is geologically the youngest of the three major provinces in the Colorado River Basin. The Basin and Range consists of fault-block mountains of metamorphic rock separated by fault-block valleys that are partially filled with erosion debris from the surrounding mountain slopes (Thornbury 1965).

The climatology of the Colorado River Basin is the product of the interaction between atmospheric systems and the earth-surface configuration (Mitchell 1976). The atmospheric system is controlled by sea surface conditions in the Pacific Ocean

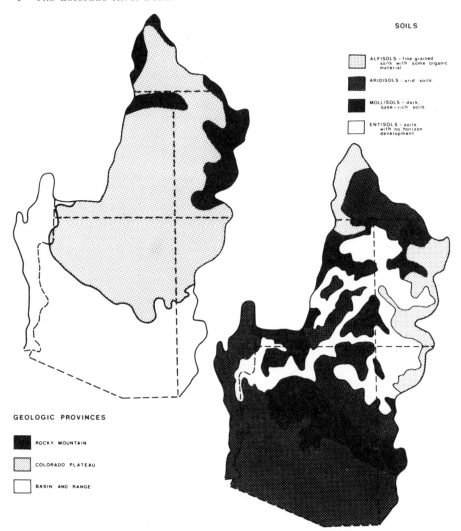

**FIGURE 2**   THE PHYSICAL GEOGRAPHY OF THE COLORADO RIVER BASIN. Redrawn from U.S. Geological Survey (1970).

which determine moisture input, and by jet stream behavior which determines the movement of fronts and pressure systems. The majority of precipitation that eventually is used by human consumers in the Colorado River Basin falls as snow in the Rocky Mountain portion of the basin or along the high altitude southern rim of the Colorado Plateau in Arizona and New Mexico. It is delivered by frontal systems or low pressure systems originating in the Pacific Ocean during the winter months. During the summer months convective storms bring intense rainfall to local areas, but these short-lived events do not significantly affect the total water supply (Sellers and Hill 1974).

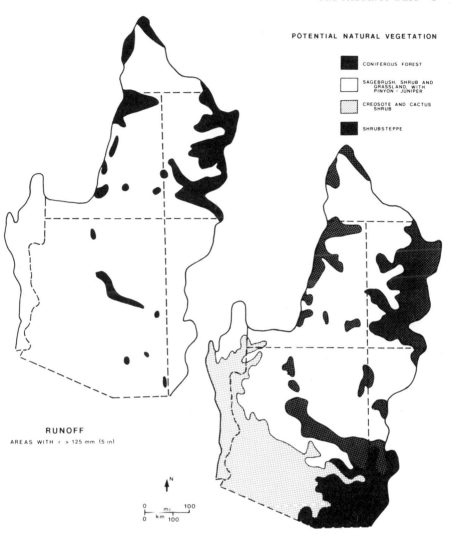

POTENTIAL NATURAL VEGETATION

CONIFEROUS FOREST

SAGEBRUSH, SHRUB AND GRASSLAND, WITH PINYON - JUNIPER

CREOSOTE AND CACTUS SHRUB

SHRUBSTEPPE

RUNOFF
AREAS WITH r > 125 mm (5 in)

N

0 — mi 100
0 — km 100

The spatial distribution of precipitation is largely controlled by orographic effects (e.g., Covington and Williams 1972). This strongly developed spatial pattern of precipitation influences the hydrologic system of the Colorado River Basin. It is only in the high mountain areas that precipitation exceeds evaporation and moisture surpluses, sufficient to generate streamflow on a continuous basis, occur. At lower elevations potential evaporation exceeds precipitation by several times, and the only continuously flowing streams are those from the mountains. The Colorado Basin therefore contains only limited source areas for surface waters, and stream discharges decline in the downstream direction in the remaining areas (Figure 2).

The combined effects of geologic materials and climatic conditions produce high variability in soils and vegetation (Hunt 1972). Although the original designations of vertically arranged vegetation zones postulated by Merriam (1890) have been frequently modified (*e.g.*, Brown *et al.* 1980), their generalizations are still useful. In the Colorado River Basin, Merriam found that at elevations below 1070 m (3,500 ft) Lower Sonoran ecological communities were dominant, wherein saguaro cactus and other arid species are common. In the Upper Sonoran Zone, between 1070 m and 1680 m (3,500 and 5,500 ft) in elevation, sage brush, pinyon pine, and juniper are the most common plants in the Colorado River Basin. Increasing precipitation and reduced evaporation that occur with increasing elevation make ponderosa pine, Douglas fir, and Rocky Mountain juniper common in the Transition Zone between 1680 and 2440 m (5,500 and 8,000 ft). In the Canadian and Hudsonian Zones ranging from 2440 to 3500 m (8,000 to 11,500 ft) water is even more abundant and lodgepole pine with aspen mix with the ponderosa pine and Douglas fir. The forests of the Transition and Canadian Zones store much of the snow for spring runoff and provided most of the local timber for the early economic development in the basin. Above these zones lies the treeless Alpine Zone, spatially limited to the highest mountain peaks in the basin.

## The Institutional Structure

A series of human, cultural, and bureaucratic systems with important spatial characteristics have administered the natural resources of the Colorado River Basin. Land ownership, for example, strongly influences the utilization of the basin resources. The United States took possession of the lands drained by the river by purchase and conquest (Robbins 1942). The majority of the basin was taken from Mexico by cession after the 1848 war, with small portions on the edges purchased from France, Mexico, and Texas (Figure 3).

During the late 1800s in an attempt to stimulate economic development, the United States federal government offered its acquisitions for sale to private owners (Rohrbach 1968). The Homestead Act of 1862 and several subsequent modifications amounted to the largest give-away of natural resources in history, but only a relatively small percentage of land in the basin passed to private ownership. The climatic conditions were too harsh to permit non-irrigated farming, and extensive irrigation depended on dam projects too expensive for private enterprise. Ranchers owned only limited lands around water sources and grazed their herds on public lands. The 1872 Mining Act permitted the extraction of minerals from public lands without requiring ownership (Wyant 1982).

Therefore, the majority of the basin is administered by the federal government, mostly in the form of Indian reservations, national parks, national forests, and Bureau of Land Management areas (Table 1; for a review of issues, see Francis and Ganzel 1984). Indian reservations are special cases because they are held in trust by the federal government but are largely administered by their native American inhabitants. Most reservations were set aside in the late 1800s and included the least desirable parts of the land from the Anglo-American perspective at that time. A century later it had become obvious that Indian lands contained valuable resources that provided their owners with a political base: coal, petroleum, uranium, timber, water, and scenery that stimulates a thriving tourist industry.

**TABLE 1**   LAND OWNERSHIP IN STATES OF THE COLORADO RIVER BASIN AS A PERCENT OF THE TOTAL AREA OF EACH STATE

| State | Private | Indian | State | Federal | FS | BLM | NPS | FWS |
|---|---|---|---|---|---|---|---|---|
| Ariz. | 28.7 | 27.2 | 13.2 | 44.1 | 15.5 | 17.3 | 2.6 | 2.1 |
| Calif. | 51.9 | 0.6 | 0.6 | 47.5 | 20.3 | 16.6 | 4.5 | — |
| Colo. | 61.6 | 1.1 | 4.5 | 37.3 | 21.6 | 12.0 | 0.8 | — |
| Nev. | 10.6 | 1.7 | — | 87.7 | 7.3 | 69.9 | 0.4 | 3.1 |
| N. Mex. | 57.6 | 9.2 | 11.9 | 33.2 | 11.9 | 16.5 | 0.3 | 0.4 |
| Utah | 32.1 | 4.3 | 6.9 | 63.6 | 15.3 | 41.9 | 1.6 | 0.2 |
| Wyo. | 48.4 | 3.0 | 5.9 | 48.6 | 14.8 | 28.5 | 3.8 | — |
| Mean | 41.6 | 6.7 | 6.1 | 51.7 | 15.2 | 29.0 | 2.0 | 0.8 |

Sources of data: Bureau of Land Management (1978) and Patric (1981). FS = Forest Service, BLM = Bureau of Land Management, NPS = National Park Service, FWS = Fish and Wildlife Service. "Federal" column is for all federal lands and exceeds the sums of FS + BLM + NPS + FWS because of other unspecified federal holdings, mostly military.

After about 1870 the spectacular landforms of the Colorado River Basin attracted the curiosity of travelers and the interest of scientists (Bartlett 1962). The result is a total of 9 national parks and 4 recreation areas (designated by Congress) plus 25 national monuments (designated by presidents) in the basin. The majority of the areas were set aside during the administration of Theodore Roosevelt, whose interest in conservation and the West also led to the designation of huge tracts of land as national forests. Under the jurisdiction of an 1891 statute (26 *U.S. Statutes,* 1103, Section 24) and later modifications, these areas were set aside for the purposes of timber production and water resource management, although since 1960 many other multiple uses have been sanctioned (Steen 1976). The remaining federal lands in the basin, except for military reservations and wildlife refuges, are literally "leftovers" that went unclaimed. Used mostly for grazing and mineral production, these lands are administered by the Bureau of Land Management (Clawson and Held 1957).

The Colorado River Basin is largely an empty quarter when compared to the nation as a whole. Despite a lack of numerous large cities, its population is one of the most highly urbanized in the country. Throughout its 629,520 sq km (244,000 sq mi) extent, the basin has only Phoenix, Las Vegas, and Tucson as cities with more than 50,000 inhabitants, yet only about 20% of the total population is rural. The basin therefore has extensive areas of extremely low population density: there are many tracts of several thousand sq km with no permanent residents (Figure 3).

Use of the land by its inhabitants is dependent on the availability of water, so the question of water rights is as important as land ownership. The states of the Colorado River Basin have differing legal interpretations regarding water rights. In the eastern and midwestern states riparian water rights prevail, where the right to use stream waters is attached to the land next to the source stream (as long as use is "reasonable"). In many riparian areas, use of the water must take place on the property next to the stream. In many western states where water is scarce water-use rights are not attached to land ownership, and the concept of prior appropriation is common. Under the precepts of prior appropriation the first user of water in time has a continuing right to the water as long as the owner uses it in a continuing beneficial manner. The right may be bought and sold separately from land parcels, and water may be transported and used far from the source. In California riparian and prior appropriation rights both occur,

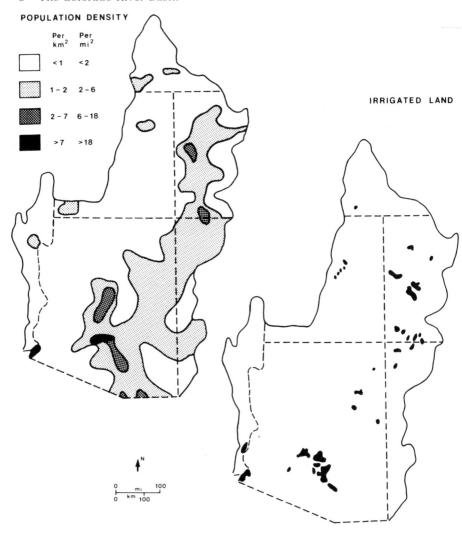

**FIGURE 3**    THE HUMAN GEOGRAPHY OF THE COLORADO RIVER BASIN.
Redrawn from U.S. Geological Survey (1970) and U.S. Bureau of
Reclamation (1946).

but in all other Colorado River Basin states prior appropriation doctrines are followed
(Trelease 1979).

The legal structure governing the use of the Colorado River must also be considered in any geographical analysis of the basin. The Colorado River Compact of 1922
provided for an initial division of the river's discharge, though the agreement was based
on faulty assumptions regarding natural conditions. In addition to the Colorado River
Compact, division and use of the basin waters are subject to a myriad of laws,

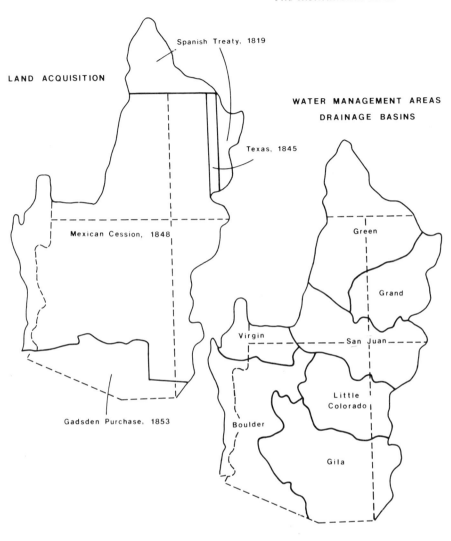

LAND ACQUISITION

Spanish Treaty, 1819

Texas, 1845

Mexican Cession, 1848

Gadsden Purchase, 1853

WATER MANAGEMENT AREAS

DRAINAGE BASINS

Green

Grand

Virgin

San Juan

Little Colorado

Boulder

Gila

agreements, and arrangements that have grown out of court cases, all collectively known as "The Law of the River" (Hundley 1966). Among the most important has been the U.S. Supreme Court case, *Arizona vs. California,* which affirmed Arizona's right to its Compact waters even though California had in place a delivery system for beneficial use while Arizona had none. *Winters vs. United States* affirmed the rights of Indians on reservations to use waters in the basin despite not being "prior appropriators" in the usual sense of the term.

Superimposed on the natural geographic patterns of the basin, the settlement pattern, and the complexities of land and water rights are the jurisdictions of three major agencies that manage the river system. The Bureau of Reclamation, formed in 1902 as part of the Department of Interior, is responsible for the construction and operation of projects that support economic development, primarily irrigation agriculture and urban-industrial uses (Robinson 1979; Warne 1973). The U.S. Army Corps of Engineers is responsible for the construction of flood control works in the basin. Finally, the Federal Emergency Management Agency, an independent federal bureau, administers the floodplain insurance programs. Each state also has its own water resource management agency, and many counties have flood control districts that further complicate the institutional structure of the basin. With so many different agencies and constituencies, consensus water policy is rare.

The interactions between society and its surrounding natural environment take place on a continuum that will be the basis of the following chapters. At one extreme along this continuum the environment is a controlling or limiting factor in human activities. Early geographic literature on this environmental deterministic perspective derived in part from supposed evidence in the Colorado River Basin which suggested that human populations adjusted by necessity in concert with climatic, hydrologic, and vegetation changes (Huntington 1914). This strictly deterministic view has been largely discredited, but there is a growing recognition of environment as hazard, a recognition that despite technological and cultural adaptations, modern society still faces occasional destructive and uncontrollable environmental processes (Burton *et al.* 1978). At the opposite end of the scale, human activities radically alter natural environments and disrupt states of equilibrium or established rates of change.

In the center of the scale of society-environment interaction lies an elusive balance where the total environment is perceived and managed by society as a resource. Some parts of the environment are intensively exploited to improve general human welfare, while other areas are left undisturbed so that preservation of some features is viewed as the best use. In the Colorado River Basin, debates over wilderness preservation versus fossil fuel development in the Canyonlands National Park area are symptomatic of the maintenance of a public policy that seeks a balanced mid-scale view of society-environment relations (U.S. National Park Services 1977d). Achievement of the balance between exploitation and preservation follows effective decisions by managers and the public they serve, decisions that depend in part on an appreciation of the environmental instability outlined in the following chapters.

# 2

# Hydrologic Instability: Flooding

One of the most destructive forms of environmental instability in the Colorado River Basin is the extreme variability in the discharge of the rivers. River floods are events where high stream flow overtops natural or artificial banks and inundates land not normally submerged (Ward 1978). Except for unusual circumstances related to earthquakes, landslides, and dam failures, river floods are the products of climatological events. The following chapter therefore begins with an exploration of the climatic systems influencing the input of precipitation to the Colorado River Basin. Channel floods and their characteristics are then reviewed, followed by a discussion of sheetfloods, a phenomenon of importance in arid and semi-arid regions. A discussion of the "artificial" floods caused by river regulation with dams concludes the chapter.

## Climatic Causes

There are three types of floods in the Colorado River Basin: rapid snowmelt, floods from long-term winter precipitation regimes, and those from short-term summer thunderstorms. Snowmelt floods occur when deep snow packs in the mountainous water surplus areas (Figure 2) melt quickly, releasing the moisture stored throughout the winter during a period of several days. The melting might be initiated by a sudden warm spell, but the more common cause is a warm rain falling on a snow pack that contains a high percentage of water (U.S. Army Corps of Engineers 1956). Snowmelt floods affect large areas of the basin at the same time.

Because the majority of the Colorado River Basin is characterized by thin soils and steep slopes, rainfall is not readily stored in the near-surface environment. When persistent winter rains saturate the meager soil layer, much of the subsequent rainfall runs off quickly, entering the stream channels as large amounts of quickflow (Dunne and Leopold 1977). The resulting large scale flooding occurs on a regional basis, affecting perhaps a third of the entire basin.

During the summer months violent convective thunderstorms are common in the entire basin. In many cases the precipitation from these storms evaporates before reaching the ground, but especially well-developed storms produce flash floods which have local impacts on basins of up to a few thousand sq km in area. Although flash floods do not affect large areas they are especially hazardous because they convert completely dry channels to rapid flow conditions within the span of a few minutes. For example, in Eldorado Canyon, a Nevada tributary to Lake Mohave, a thunderstorm in

1974 converted the dry canyon floor to a river with a peak discharge of 2,100 cubic meters per second (cms, or about 76,000 cubic feet per second, cfs) in a matter of minutes, killing 9 people. Although the stream flowed for only three hours, it carried 2.5 million cu m (2,000 acre feet) of water and 17,600 cu m (70,000 cu yd) of sediment into the lake (Glancy and Harmsen 1975).

An appreciation of the causes behind each of the flood types requires identification of the source of moisture for the floods and specification of the climatic mechanism by which the moisture is transported into the basin (Figure 4). There appear to be three sources of moisture that affect the Colorado River Basin: the east-central Pacific Ocean, the Gulf of Mexico, and the Gulf of Baja California (Hales 1974). The source of the majority of moisture entering the basin is the Pacific Ocean, where evaporation from the surface injects moisture into the generally westerly flow of air. Shifts in ocean currents and changes in general cloudiness alter the temperature of the sea surface from time to time, which in turn changes the rate of evaporation and the rate of moisture supply to the basin from this source (Harnack and Broccoli 1979; Namias 1969). When ocean surface temperatures are relatively warm, the evaporation process accelerates and the basin receives increased amounts of precipitation.

A second source of moisture is the surface of the Gulf of Mexico which injects moisture into easterly winds that sweep across Texas, New Mexico, and northern Mexico. Moisture for summer thunderstorms sometimes follows this track. Finally, the Gulf of Baja California, between the Baja Peninsula south of California and the mainland of Mexico, supplies moisture for the "summer monsoon," a period of relative high humidity (30-40%) in the Colorado River Basin during the July-September period (Hales 1974). Afternoon thunderstorms generated by convection driven by high temperatures (in excess of 40°C or 100°F) on and near the desert surface convert the moisture to precipitation.

The frequency with which moisture from these three sources enters the Colorado River Basin and the intensity of the storms which convert it to precipitation are direclty related to the location of the polar jet stream over the western United States. This fast-moving river of air flows from west to east at 9,150-12,200 m (30,000-40,000 ft) above the surface. It has a meandering course that circles the globe in northern latitudes, usually entering the continent during the winter at the coast of Washington or Oregon. It then loops southward to the central Great Plains, and then northward through the Great Lakes region, exiting the continent across the New England coast. Storm systems in the form of low pressure cells and fronts are commonly aligned along the jet stream (Harman 1971), so a typical winter condition is for the storms to be steered on-shore in Washington and Oregon and then into the continental mid-section, largely by-passing the Colorado River Basin except for its northern extremities. During most years with this "normal" jet stream orientation, snowmelt and winter floods are uncommon (Sellers and Hill 1974).

Approximately every other decade (1900s, 20s, 40s, 60s, and 80s), the "normal" pattern breaks down, however, and is replaced by a new jet stream pattern in which the stream is displaced southward in the western United States. Entering the continent in central or southern California during these brief periods, the jet stream steers winter low pressure systems across the warm east-central Pacific directly into the Colorado River Basin and produces heavy rains at low elevations with deep snow packs in the mountains (Barry et al. 1981). If the pattern persists for several weeks as it did on several occasions during the period 1978-1983, flooding is common and water supplies are likely to be especially abundant.

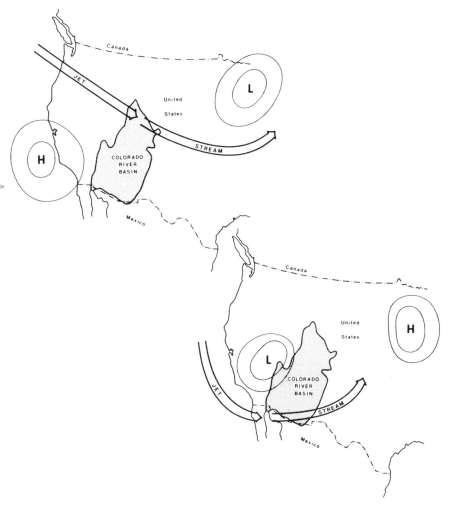

**FIGURE 4** CLIMATIC PATTERNS AFFECTING THE COLORADO RIVER BASIN. Above: Circulation pattern characteristic of relatively dry winter conditions in the Colorado River Basin that results in sparse water supplies. Below: Circulation pattern characteristic of moist winter conditions in the basin that result in floods and plentiful water supplies.

The significance of the variability of this atmospheric moisture delivery system is that water resource managers must expect long periods with few floods and below average surface water supplies interspersed with brief periods of high runoff. In the short period of instrumented records before the Colorado River Compact of 1922, two periods of dominance by southerly positioned jet streams resulted in runoff that apparently was far above the average on a multi-decade or multi-century basis (Figure 5; Stockton 1975). Managers of floodplains in the basin were lulled into a false sense of security during the period between the early 1940s and the middle 1960s when jetstream positions were "normal" and flooding was rare. When southerly shifts became common in more recent times, floods and floodplain damages were especially acute. Planning for variability rather than for averages is therefore the most likely route to success.

## Channel Flooding

Channel flooding occurs when the channel defined by banks is not large enough to accommodate the discharge provided by surface runoff. In the Colorado River Basin, channels have different physical characteristics than those in the eastern and mid-western United States where flood protection technology and planning were developed. Flood problems in the Colorado River Basin derive as much from this mismatch between problem and solution as from the seemingly unpredictable nature of the climate. Channels in the basin are primarily of four types: main rivers, secondary streams, channels on alluvial fans, and channels on basin floors.

The main rivers of the basin, the Green, Colorado, and San Juan, have tremendous variability of flow on a seasonal basis. In their steep upper reaches these rivers are mountain streams with rocky channels and few floodplains. Lower down, the rivers flow through broad valleys with very wide floodplains used for irrigation agriculture. In the middle of the basin all three rivers flow through steep rock-bound canyons with hardly enough room on the floor for the channel. The lower Colorado River, south of Hoover Dam, has a relatively gentle gradient and has a broad irrigated floodplain. In the steep, restricted portion of the channel, variability of flow affects little land, while in the low-gradient sections flooding over large areas and nearly dry channels may both be experienced in the same year. Unlike eastern or midwestern rivers there are no smooth spatial or temporal transitions along the stream.

Smaller secondary channels, tributary to the main rivers, are frequently braided in order to accommodate high seasonal variability of discharge, large amounts of sediment from the surrounding arid and semi-arid landscape, and poorly consolidated bank materials (Miall 1977). During the spring melt period these streams are wide and shallow, but during the remaining low-flow periods of the year they consist of only a single low-flow channel accompanied by numerous dry channels. Banks are frequently indistinct and definition of a floodplain is virtually impossible due to changes in channel shape, size, and pattern that occur during every flood.

In the Basin and Range portion of the Colorado River Basin alluvial fans are economically important landscape components because they are frequently locations of groundwater supplies and provide valuable space for urban development (Bull 1977). Each alluvial fan is built by sediment delivered by a single channel issuing from a mountain front, but that single channel disintegrates into a series of distributary channels across the fan surface. These small channels are especially difficult to control because they are shallow and unstable; all channels are not active at any one time. The smallest channels may be eliminated by development activities, causing the remaining channels to carry unnaturally large amounts of water with resulting erosion and sedimentation problems (Cooke *et al.* 1982).

As medium and small ephemeral channels issue from mountains or mesas onto relatively flat valley floors or bolsons, they lose definition and under natural conditions their flood waters spread out in shallow flows as much as 5 km (3 mi) wide. When the valley floors are developed for irrigation agriculture the flow zones are frequently not

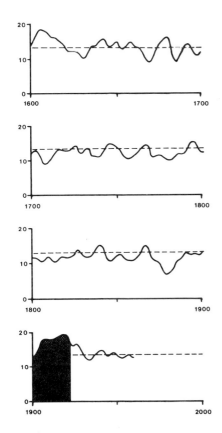

**Virgin Discharge At Lee's Ferry, Arizona (10⁶ Ac. Ft.)**

**FIGURE 5**    LONG TERM RECONSTRUCTED DISCHARGE RECORDS FOR THE COLORADO RIVER SYSTEM BASED ON TREE-RING INDICATORS. Dashed line represents the mean value of 13 million acre feet per year. The shaded portion of the diagram represents the period used for the estimates in the Colorado River Compact of 1922, clearly an anomolous period. Redrawn from Stockton (1975).

recognized since they may have no banks. When floods occur, wide areas are affected. The usual remedy is to install drains and artificial channels, but these features usually depart so widely from the natural flow zones that destructive erosion and sedimentation result.

Queen Creek in central Arizona provides an instructive example of the management problems associated with economic development near ephemeral streams in the Colorado River Basin. Under natural conditions Queen Creek rose in the mountains of east-central Arizona, flowed westward out of the mountains, and spread out over the deep alluvial fills of the Salt River Valley. Early irrigation survey maps show that in 1902 the channel became progressively smaller as it crossed the valley floor and lost its waters to percolation and evaporation. The stream flowed in a wide zone to the Gila River only during periodic floods.

During the period of World War I high grain prices stimulated dry-land farming of the area on the valley floor in the flow zone and even in the consistently dry channel. In 1925 a canal was constructed *across* the alignment of the channel and flow zone, attesting to its indistinct nature. When the upper watershed of Queen Creek received heavy rainfall, the lower creek flooded, overtopping the canal and destroying fields. Later, urban development on the valley floor sent increased amounts of storm runoff into the system, increasing the problems of flood control. Occasionally, a path of post-flood destruction marks the old natural flow zone of the stream that defies management techniques derived in humid regions.

## Sheetflooding

When raindrops fall on the ground surface and fail to percolate into the soil they begin to collect in shallow depressions and eventually the water mass moves down-slope. At this stage of the runoff process the water moves as a shallow sheet, not yet collected into rills or channels (Dunne and Leopold 1977). Even as a thin sheet the water entrains soil particles and carries them downslope, causing some erosion. Eventually the flow concentrates into drainage lines formed by small rills, and the flow processes including sediment transport become channel phenomena. That these processes of sheetflow and sheet erosion occur in some humid regions is clear, but their role in sheetfloods in arid regions is the subject of considerable debate.

As federal surveyors and scientists explored the western United States generally and the Basin and Range portion of the Colorado River Basin in particular, they found that hillslopes had geometric configurations that were radically different from the shapes of hillslopes in the more familiar east and midwest (Bartlett 1962). Broad aprons of coalesced alluvial fans and long, gently inclined pediments connected many mountain ranges with the adjacent flat surfaces of alluvial valley fills. McGee (1897) proposed that the surfaces of alluvial fans and particularly pediments were the product of sheet floods, sheets of runoff that shaped the land surface after intense rainfalls. He reported on the physical appearance of such a sheetflood which he observed while he was horseback riding in southern Arizona.

Subsequent observers and authors debated the role of sheetfloods in arid regions, but no one reported witnessing the actual process, a situation which naturally led to some confusion (Rahn 1967). It was obvious that sheetflow in arid regions on the footslopes of mountains was not the same as sheetflow just below the crest of a gently sloping hill in humid regions. Davis (1938) pointed out that discharges that looked like the McGee sheetflood occurred in arid regions where numerous small channels on alluvial fans overflow and combine. Also, wide shallow sheets of rapidly moving water develop when a channel empties its flood waters onto a relatively flat alluvial valley as described in the example of Queen Creek.

Whatever their cause and genesis, sheetfloods do occur in the Basin and Range portion of the Colorado River Basin and are a source of considerable hazard to economic development. Irrigation agriculture, for example, depends on an intricate system of major feeder canals and minor lateral canals, all delicately adjusted to each other in terms of gradient. When sheetfloods wash over the canals the channels are filled with silt from nearby fields and may be structurally destroyed. The surfaces of the fields are carefully contoured and smoothed so that a precise gradient exists from the entry point for irrigation water and the farthest corner of the field. This "leveling" is accomplished with laser-guided tractors, but sheetflooding disrupts the planned surface by eroding some parts and depositing sediment in other areas, requiring the complete reconstruction of the field.

Sheetfloods from undeveloped lands also pose problems for expanding urban areas (for a general review of urban flood problems, see Bauman and Kates 1972). Waters collect on the undeveloped land and flow in sheets down to developed areas that already suffer from insufficient drainage to collect runoff from nearby urban surfaces during intense rainstorms. The sheetfloods arrive in a dispersed fashion rather than concentrated in a single flow thread, so they first must be collected from a wide area before disposal through a drain system. Highways, railroads, airports, and long distance canals are frequently built on gentle pediment slopes and must be protected from sheetfloods by expensive control works.

Examples of the problems and solutions associated with sheetfloods are provided by the Colorado River Aqueduct and the city of Scottsdale, Arizona. The Colorado River Aqueduct carries Colorado River water from a pumping station on the shore of Lake Havasu behind Parker Dam to the cities of Los Angeles and San Diego. West of the Colorado River the aqueduct crosses the lower portions of coalesced alluvial fans. Upslope from the aqueduct long, low walls, or wings, angle away from bridge crossings over the canal. These walls, which taken together appear like chevrons with the bridges at their downslope apexes, collect the sheetflow into a manageable small area, and then the bridges conduct the flow across the aqueduct without damage.

The city of Scottsdale, a rapidly expanding suburb of Phoenix located near the foot of the McDowell Mountains, provides another example. As the residential districts of the city encroach on the pediments and fans near the mountains, streamflows from the undeveloped lands upslope are less concentrated and more commonly take on the appearance of sheetflow (Rhoads 1984). The few channels that exist are unstable and provide inadequate capacity for water from the sheetfloods if it were to be collected for more easy handling by drains. The city government has been forced to abandon traditional planning and engineering techniques and to consider prohibiting development near large drainways that may conduct the partially concentrated sheetfloods.

## "Artificial" Floods

The Colorado River system has been called the world's largest plumbing system (Fradkin 1981) because of the degree of control on river flow exerted by its numerous high dams. As stated earlier, one of the purposes for constructing the dams was flood protection for valuable agricultural lands on the lower reaches of the river in California and Arizona (U.S. Army Corps of Engineers and U.S. Bureau of Reclamation 1982). Generally, high discharges occur on these reaches of the river only if upstream operators open floodgates and permit large quantities of water to pass the dams. Recognizing this supposed control capability, downstream residents who found their homes accessible only by boat through the second floor windows in 1983 charged that they were plagued by an "artificial" flood resulting from negligence on the part of the dam operators.

Reservoir management for the complex multi-structure Colorado system is a tricky business, however, and understanding of the "artificial" floods requires an appreciation of the broad concepts of flow regulation, multiple use, and discharge prediction (Rutter and Engstrom 1964). Flow regulation refers to the capability of a large dam and associated reservoir to accept highly irregular discharges as input and to release a highly controlled flow as output. In order to accomplish this river control, arid-region reservoirs must have large capacities for storage because unlike their humid-region counterparts they must receive the majority of the annual input within only two or three months. The irregularities of input are absorbed by the storage in the reservoir, so that during periods of high input, most is stored, while during dry periods the storage component is drawn down by releases. Most of the large reservoirs in the Colorado River system were designed to store about two years of continuous input, because not only are the seasonal fluctuations great, but there are also radical fluctuations from one year to the next.

Reservoir management is complicated by multiple objectives, however. The case described above is for relatively constant output from the reservoir, a situation desirable for downstream navigation. Most dams, including those of the Colorado River system, must address other needs as well. Agricultural irrigation has heavy water demands during mid- and late summer when crops are most actively developing and when evapotranspiration is highest. Low demands in winter result from little crop activity and low evapotranspiration. Urban users also have highest demands for water in summer to meet cooling needs (Martin *et al.* 1984). The generation of electricity requires diurnal fluctuations in releases through the dams and their turbines, with peak periods in the early evening hours. Recreation interests prefer to have the reservoir pool at the highest possible level all the time to provide the potential for water sports. Occasional demands for flood protection result in the need to draw down the pool prior to the expected flood season to accommodate potentially hazardous inflows. Faced with these competing demands, reservoir managers attempt to strike some balance, but all users are not likely to be equally satisfied.

Finally, there is the problem of prediction. A successful multiple use strategy depends on the knowledge of how much inflow to expect over the course of a given year. In the case of the Colorado River Basin this resolves into the need to accurately assess inflow from the spring snowmelt. The amount of water expected from this source depends on the amount of snow stored in the upper watershed (ultimately dependent on the climatic delivery systems for moisture), its moisture content, and subsequent spring rains (U.S. Army Corps of Engineers 1956). The snow pack is regularly measured and sampled throughout the winter by ski-borne hydrologists and self-operated measurement stations that beam periodic reports to the SNOTEL satellite network. Even with these fingers on the snow-hydrology pulse, reservoir operators cannot assess the impacts of unpredictable climatic events such as a sudden warming trend or a surprise late winter storm.

The 1983 flooding along the lower Colorado affected communities in California, Arizona, and Mexico. Under the original pre-dam conditions floods were an annual occurrence along the lower reaches of the river, and communities were constructed on low bluffs overlooking the channel and floodplain. When Boulder Dam (later called Hoover Dam) was constructed in the 1930s, part of its justification was as a flood control structure to protect the lower Colorado River. The dam, along with subsequent structures, was remarkably successful in eliminating the devastating floods that once plagued the area. Urban development in the formerly recognized floodways followed.

The complex multiple use strategies and problems of prediction finally brought about the "artificial" floods of 1983. When unexpected late winter snows melted rapidly the reservoirs throughout the Colorado River Basin had inadequate storage capacity to absorb the inflow, and in order to avoid destructive flows over the tops of the structures, large quantities of water had to be released. Parker Dam which had previously not released more than 700 cms (cubic meters per second; 25,000 cfs, cubic feet per second) ultimately had to release more than 1,260 cms (45,000 cfs) for several weeks in 1983. The result was that parts of several riverside communities were under water.

The long-term successful operation of the dams lulled property users near the river into a false sense of security. Failing to recognize the long-term variability of climatic conditions and the multiple objectives of the upstream reservoir operations, they built homes and businesses in hazardous locations. Seeking scenic views of the river, they built on floodplains even when local zoning ordinances prohibited the practice. Seeking inexpensive land they built homes in front of levees, between the protection works and the river where high waters flowed unchecked. Through private development a few took the risks, but by public disaster aid, society paid the cost.

The Colorado River Basin, therefore, is an example of most arid and semi-arid regions in that its hydrologic system is subject to great instability and variability. Planning and management of the basin's resources have been most successful when society's approach to the region has taken this variability into account (Costa 1978). When society has relied on short-term averages for decision-making, the results have been legal, economic, and social problems with unwise development in hazardous locations. Structural controls of flows must be augmented by non-structural efforts such as hazard zoning in spite of the attending legal and social costs (Dingman and Platt 1977). Instead of unusual events, floods in arid regions such as the Colorado River Basin are an expected component of a complex, unstable environmental system.

# 3

# Ecological Instability: Riparian Vegetation

The variability of hydrologic systems in the Colorado River Basin is made more difficult to manage by attending instability in the ecological systems near the stream channels. These riparian plant communities are dynamic even in undisturbed situations, but with the advent of economic development in the basin the ecological instability of the plant communities has increased (Johnson 1970). The purpose of the following chapter is to briefly explore the significance of the riparian plant community for human activities, to explain how the native community has changed as a result of the invasion of exotic species, and to outline the scientific and management problems associated with the relationship between water resources and the riparian vegetation.

## Nature of Riparian Vegetation

Riparian vegetation consists of those plants growing in and near river channels and directly influenced by river-related processes. Riparian vegetation interrupts the model of altitudinal variation in vegetation communities described in Chapter 1 because the riparian plants depend on river conditions as well as altitude for the definition of their range (Pase and Layser 1977). In upper elevations in the Colorado River Basin, the riparian community is dominated by willow (*Salix goodingii*), but at lower elevations the community is more diverse. In the Transition, Upper Sonoran, and Lower Sonoran vegetation zones the natural riparian community consists of mixtures of willow, cottonwood (*Populus fremontii*), mesquite (*Prosopis* ssp.), seepwillow (*Baccharis glutinosa*), and arrowweed (*Pluchea*). Many riparian species are phreatophytes, plants with extensive taproot systems that allow them to draw moisture directly from the saturated zone below the water table.

The significance of riparian vegetation from the societal perspective includes its value in preserving ecological diversity, landscape aesthetics, and wildlife management; its problems related to flood control and sedimentation; and its mixed impact on the management of water resources (Robinson 1958; Horton 1972). Because of its special location, riparian vegetation is frequently different than the vegetation communities that surround it, so that any activities that destroy riparian vegetation seriously affect the diversity of the entire vegetation community. Throughout the

Colorado River Basin the diversion of streamflow for irrigation has caused the blossoming of productive agricultural areas, but it frequently has destroyed the natural riparian vegetation downstream from the diversions by altering the magnitude and duration of streamflow. In Arizona only about 15% of the natural riparian communities that once existed have survived the interruption of natural streamflows for irrigation (Brown *et al.* 1977). On large rivers, high dams and their regulated outflows may have created an environment more favorable for riparian vegetation (Turner and Karpiscak 1980).

The loss of riparian vegetation has had an important impact on the appearance of the landscape (Graf 1979a). In desert regions such as those that dominate the Colorado River Basin, the overwhelming impression is one of stark landforms and (during summer) one of bleak brown vistas. Riparian vegetation introduces a welcome relief in the form of a relatively lush growth of green that is a pleasant contrast to the surrounding harsh desert. Residential areas near riparian zones have higher property values reflecting the positive externality offered by the vegetation.

The distribution of animals is closely related to the distribution of plants, so the riparian communities in arid regions are especially important as wildlife habitats (Anderson *et al.* 1977). Species diversity, especially for birds, is enhanced in the riparian vegetation zones, and such areas are important to certain specific species (Cohan *et al.* 1978). The white-winged dove, a popular game bird, favors riparian zones for breeding, especially if agricultural food sources are close by (Arnold 1973). In many parts of the Colorado River Basin bald eagles (an endangered species) nest in cottonwood trees which provide secure bases from which small game and fish may be obtained for the young (Fish and Wildlife Service 1976). The elimination and alteration of riparian habitats have direct impacts on wildlife populations.

Riparian vegetation management is an important component of flood-control measures because the densely growing plants obstruct the free movement of water across floodplain surfaces and through channels where they grow (Graf 1980a). Because phreatophytes survive primarily on water drawn from the subsurface, they may thrive in an ephemeral channel which has been colonized during a period when there is no surface flow. When a large discharge occurs, the channel may be choked with phreatophyte vegetation which reduces channel capacity, forcing the water onto nearby lands. A flood results in an area where it would not have occurred without the phreatophytes.

During flood events, large amounts of sediment are carried by the stream flow, but riparian vegetation causes sedimentation to occur (Hadley 1961). As the water flows around stems and low branches, turbulence develops which lessens the ability of the flow to transport sedimentary particles. The particles are deposited in sizeable accumulations around the plants, causing further reductions in channel capacity which increases the likelihood of destructive flooding.

Because riparian vegetation uses water for its growth and development, some water resource managers in the Colorado River Basin have favored its elimination as a means of increasing the amount of water available for human use (Gatewood *et al.* 1950; Culler 1970). These "water salvage" efforts are usually directed at concentrations of phreatophytes because they have been perceived as the most prodigious water users. Clearing efforts have occurred along many major streams in Arizona, Utah, and New Mexico, sometimes with the additional objective of establishing vegetation-free floodways (Horton and Campbell 1974).

## Exotic Species

An exotic species is one which is introduced by human activity to a region where the species is not naturally found (Clark 1956). Flowering plants are frequently introduced for gardening and ornamental purposes, but exotic species usually escape cultivation and compete with native vegetation for space in the natural environment. Usually the native vegetation is successful against the exotic plants because the native species have evolved to fill particular ecological niches. Sometimes, however, the exotic is able to compete successfully against the original natural vegetation. A successful invading species can cause radical alteration and instability in the original natural vegetation community and may virtually eliminate the original species. In the Colorado River Basin two exotic species, tamarisk and russian olive, have caused profound changes in the riparian communities and have been the subject of millions of dollars in management efforts (Harris 1966).

Tamarisk (*Tamarix chinensis,* Lour., also known as salt cedar) is a phreatophyte that commonly grows as a shrub or tree with wispy, scale-like branches and fragrant lavender blossoms (Baum 1967). A native of the Mediterranean region, it was imported into the United States as part of a seed exchange program and was sold in California nurseries in 1852 as an ornamental plant and shade tree (McClintock 1951). The taxonomy of the plant is confusing, but there appear to be several varieties, with the desirable types being those that grow as single-trunked trees (Horton 1962). Some seeds were brought to the United States that produced a smaller, more shrub-like plant, however, and when these plants escaped cultivation sometime in the late 1800s they aggressively invaded riparian environments throughout the American West, especially in the Colorado River Basin (Robinson 1965; Horton 1977).

Tamarisk seedlings are air-borne, quickly take root on moist sand surfaces, and grow rapidly — as much as 3 m (9 ft) per year (Horton 1964). These characteristics allowed it to compete effectively with native vegetation along the rivers of the basin (Horton *et al.* 1960). Because of the extreme seasonal variation in discharge, the major rivers expose large areas of moist sand bars in the late spring and early summer as discharges decline after the snowmelt. This time period also coincides with the time of maximum seed production for tamarisk (Warren and Turner 1975). Rapid growth rates and extremely effective taproot systems allow the invader to crowd out the native vegetation, mostly willow and cottonwood. In addition, tamarisk was intentionally planted in New Mexico in the late 1800s as an erosion control measure with the hopes that it would stabilize stream banks. In Arizona it was viewed as a desirable ornamental that was planted around farmsteads.

The initial points of introduction of tamarisk into the natural riparian environment are not clear, but the earliest evidence of its growth in uncultivated circumstances is an 1892 photograph of the Salt River near Phoenix which shows a tamarisk seedling (Graf 1982a). It was established in the Virgin River System in southern Utah and Nevada during the 1890s (Bowser 1960), but the plant was not common in the basin before 1900. By 1920, however, tamarisk occurred in most drainages throughout the basin (Clover and Jotter 1944; Harris 1966), and dramatic spread was noted by several observers during the post-1935 period (Christensen 1962). It appears that tamarisk was able to withstand drought effects such as those of the mid-1930s more effectively than native vegetation, so that when moist periods occurred in the 1940s it was in a dominant competitive position. Generally, tamarisk spread from the southern portion of the basin northward at an average rate of about 20 km (12 mi) per year (Graf 1978).

In 1920 tamarisk covered about 4,000 hectares (10,000 acres) in the western United States; in 1961, 364,000 hectares (900,000 acres), and in 1970, 526,000 hectares (1,300,000 acres; Robinson 1965). Despite supposed restrictions on its range (Harris 1966), the plant was found from the Colorado Delta in Mexico to small streams high in the Rocky Mountains. Its most profound impact was in the central and lower portions of the basin, however, where it colonized the entire floodplain of medium and large rivers as well as channel areas (Turner 1974). Dense thickets of tamarisk had developed along the Colorado, Green, San Juan, Virgin, Gila, and Salt rivers where once bare sand bars or modest growth of willow and cottonwood had prevailed (Figure 6).

Even the resilient tamarisk could not escape the consequences of changing environmental conditions. The extent of its coverage and the density of its growth were subject to the influences of surface and groundwater (Horton 1977; Graf 1982a). When dams were constructed and water releases through channels severely curtailed, the tamarisk communities downstream from the dams withered and declined. In the reservoir areas created by the dams, however, tamarisk growth expanded as new surfaces of moist sand were exposed by fluctuating levels of the lake waters. On balance, there probably was more tamarisk after the closure of dams than before. Some authors have even suggested that the spread of tamarisk was caused by the construction of dams (Harris 1966), but such an outcome seems unlikely in view of the successful invasion of the species prior to construction and its continued spread in streams without dams.

The taproots of tamarisk are not able to completely insulate the plants from the effects of changing groundwater levels. If the groundwater levels decline slowly, the roots can grow rapidly enough to keep pace and to continue to supply the plants with moisture (Gatewood *et al.* 1950). When the water table is deeper than about 10 m (33 ft), however, dense growth of tamarisk no longer survives. Along the Salt River in the Phoenix area, dense thickets of tamarisk were converted to barren cobble and sand surfaces when pumping by wells drew the water table down to depths greater than 10 m (33 ft). Dense tamarisk has not returned as the water table has continued to decline to depths of greater than 100 m (330 ft).

## Riparian Vegetation and Water Management

The connection between phreatophyte growth and groundwater has posed one of the most long-standing and expensive environmental management problems in the Colorado River Basin (Robinson 1965). Groundwater has been critical to the agricultural and urban development of the basin because of the scarcity and variability of surface water supplies. In the central portion of the basin most cities and farmers supplement their surface water supplies with pumped well water, but in the more arid southern part of the basin groundwater literally makes economic activities possible. Las Vegas, Tucson, and many smaller cities in desert settings depend almost completely on pumped groundwater, while the Phoenix urban area obtains about 40% of its water from subsurface sources. Any use of groundwater other than for clear economic gain is therefore subject to careful management in hopes of reducing "losses." These efforts are particularly important because in most arid areas of the basin human withdrawal and use of groundwater exceeds the rate of natural replacement. Water

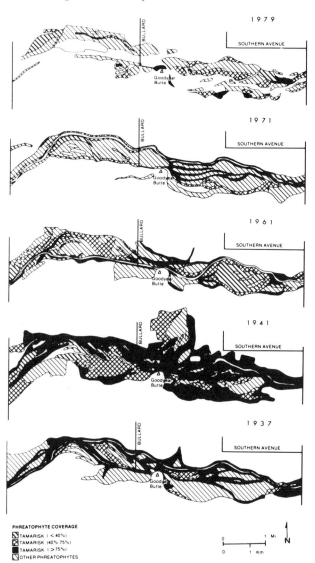

PHREATOPHYTE COVERAGE
- TAMARISK ( < 40%)
- TAMARISK (40% - 75%)
- TAMARISK ( > 75%)
- OTHER PHREATOPHYTES

0        1   Mi
0    1   Km
N

**FIGURE 6**     THE CHANGING DENSITY AND COVERAGE OF PHREATOPHYTES ALONG THE GILA RIVER IMMEDIATELY WEST OF PHOENIX, ARIZONA. The 1937-1941 period represents the time of maximum development of the phreatophyte community as the tamarisk invasion reached its height. By 1961 pumping of groundwater for agriculture had caused a drastic decline in the water table and the phreatophytes were declining. In 1971 the dense growth of phreatophytes was limited to areas near the minor low-flow channel and ribbons along channels shaped by flooding in 1965-1966 that in 1971 drained local irrigated fields. In 1979 the pumping continued to cause water table declines and floods had destroyed additional phreatophyte areas. Data from interpretation of aerial photographs.

used in the twentieth century may have entered the system several million years ago and is not likely to be replaced in the foreseeable future.

In most arid-region basins the water table is closest to the surface near streams because the channels lose water to the subsurface (Figure 7). This mound of groundwater near channels is tapped by the roots of phreatophytes which then use the groundwater for growth. During the accelerated economic activities in the mining and agricultural sectors during World War II, water in the Colorado River Basin became increasingly important, and some investigators proposed the "salvage" of groundwater by eliminating the phreatophytes (Gatewood *et al.* 1950). The result was a 40 year debate among scientists and water managers revolving around the central issue of how much water would be salvaged by phreatophyte removal (Hughes 1971).

The significance of this question is that if very little water can be salvaged the cost of phreatophyte removal will exceed the benefits and the effort will not be worthwhile (Culler 1970). If large amounts of water can be saved, then some environmental costs in addition to the costs of removing the plants would probably be acceptable. In order to make a reasonable decision, managers therefore need to know how much moisture is being transpired by the phreatophytes. This same question is also of interest to hydrologists who want to know the relative role of phreatophytes in arid-region water balance studies.

The measurement of water use by phreatophytes (and hence, potential water salvage) turned out not to be a simple matter (Graf *et al.* 1984). The complexity of natural hydrologic systems, difficult field conditions, correlation of field and laboratory results, and variable hydro-climatic conditions have resulted in a confused body of literature produced by scientists and engineers that provides little useful guidance for the decision makers. A brief review of the investigations into the question of phreatophyte evapotranspiration and potential water salvage is a useful object lesson in the limitations of environmental science.

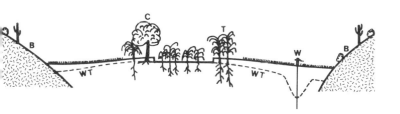

**FIGURE 7**    A CROSS SECTION OF A TYPICAL ARID-REGION RIVER VALLEY WITH ALLUVIAL FILL AND A WATER TABLE INTERSECTING THE SURFACE AT THE RIVER CHANNEL. **B,** bedrock valley sides with xeric vegetation. **C,** cottonwood tree on stream bank wth moderate root system. **G,** grass-covered floodplain. **T,** tamarisk trees and shrubs growing on stream bank and in the channel with extensive root systems. **W,** an artificial well pumping water from the saturated zone. **WT,** the water table with a cone of depression around the well.

Except for some experimental electronic methods (Gay and Hartman 1981, 1982; Erie *et al.* 1982), measurement of phreatophyte transpiration has generally taken one of three routes: lysimeters, tent techniques, and water balance approaches. Lysimeters are large tanks containing plants and their associated soil (van Hylckama 1974). By weighing the tanks when water is added, water loss through evapotranspiration can be estimated. Lysimeters provide direct measures of water in highly controlled environments, but their physical limitations prevent simulation of the real, complex environment. More evapotranspiration studies have been accomplished with lysimeters than any other approach, but comparison of results is confused by variability in sizes of tanks, plant types used, and experimental designs that have ranged from complex plant assemblages in some cases to the use of one plant per tank in other studies.

In attempts to study more accurately the conditions in the field, some workers have employed plastic tents that completely encased one or more plants *in situ* (Sebenick and Thames 1967; Tomanek and Ziegler 1961). In this approach, air is pumped through the tent and the humidity of the input is compared with that of the output. The difference is assumed to be the result of evapotranspiration. The major advantage of the tent technique is that it provides direct measurements of evapotranspiration in a relatively undisturbed natural setting. The disadvantages are that the measurements can be made for only one or at most a few plants at a time rather than for an entire community, and that the length of the measurement period is relatively brief. The method cannot effectively account for temporal variability.

At the opposite end of the scale of analysis is the water balance approach (introduced by Thornthwaite and Mather 1955, 1957). In this method all the water inputs and outputs of a river reach or drainage basin are accounted for so that the relative significance and magnitude of evapotranspiration can be assessed through data collected from wells, stream gauges, and precipitation gauges over a period of many months. The water balance approach is useful because it represents a whole-system, integrative analysis that accounts for regional complexity on a time scale long enough to account for some temporal variability. The imperfections of the water balance approach include the inability to measure inputs and outputs accurately in uncontrolled field conditions and the relatively small proportion of the total water budget accounted for by the object of the study: evapotranspiration. If errors occur in the measurement of other components of the water balance they seriously affect the evapotranspiration estimates.

The results of nearly four decades of evapotranspiration measurements and research for the Colorado River Basin and nearby similar basins do not provide much in the way of specific guidance for decision makers (Graf *et al.* 1984). As shown in Table 2 there is great variability in reported rates of evapotranspiration for tamarisk, the phreatophyte of most concern. Some of the data are not useful for comparative purposes because they are reported in nonstandard units, necessitated by the nature of the experimental designs. It is difficult, if not impossible, to make an informed opinion about evapotranspiration and potential water salvage when estimates of annual rates range from 34 cm (13.2 in) to 280 cm (110.0 in).

## Clearing of Riparian Vegetation

Despite uncertainty about the potential amount of water savings through the elimination of phreatophytes, millions of dollars have been spent on clearing projects

**TABLE 2**  MEASURED EVAPOTRANSPIRATION RATES
REPORTED FOR PHREATOPHYTES AND OTHER RIPARIAN
VEGETATION IN THE SOUTHWESTERN UNITED STATES

| Species | Rate (cm water per year) |
|---|---|
| Bare Ground | 64-81 |
| **Phreatophytes** | |
| Cottonwood | 154-248 |
| Mesquite | 101.4 |
| Tamarisk | 34-280 |
| Willow | 77-139 |
| **Shrubs** | |
| Arrowweed | 244 |
| Fourwing Saltbush | 109 |
| **Grasses** | |
| Bermuda | 107-276 |
| Blue Panic | 124-133 |
| Alta Fescue | 182 |
| St. Augustine | 166 |
| Alkalai Sacaton | 43 |
| Saltgrass | 54 |
| **Crops** | |
| Alfalfa | 175-213 |
| Barley | 30-112 |
| Cotton | 65-105 |
| Wheat | 66-112 |

Data from Patten (1984).

along rivers in the southwestern United States (Johnson 1970). These efforts at active manipulation of the riparian ecosystem have been focused on tamarisk but have failed to produce the expected water savings and have revealed two major management problems for areas that are cleared: maintenance and replacement vegetation.

Once tamarisk has been cleared from a floodplain or section of channel, the problem unfortunately has not been solved (Horton and Campbell 1974). Clearing produces bare sandy surfaces that are likely to be moist at some time of the year because of their riverine location; this of course presents prime seed beds for the re-establishment of tamarisk seedlings. Propagation by suckers on stems and roots of cut plants, production of hundreds of thousands of air-borne seeds by surviving plants, and rapid growth rates have insured that those areas cleared of phreatophytes have been recolonized by dense growth within 18 to 24 months of the clearing effort. Periodic maintenance of the cleared channel is therefore required, significantly adding to the cost of the salvaged water. The maintenance efforts may take a variety of forms, including mowing, cutting, burning, drowning, or poisoning, but all have undesirable environmental side effects ranging from air and water pollution to loss of wildlife habitat and the production of unsightly cut-over areas (Hughes 1971).

A potential solution to the continuing maintenance problem is the installation of replacement vegetation. Unfortunately, the planting and nurturing of water-conserving grasses to replace phreatophytes is an expensive proposition that would demand most of the water originally saved by removal of the phreatophytes. Continuing maintenance is not avoided because the grasses must be managed through periodic replanting.

The Gila River, a principal lower-basin tributary to the Colorado River which rises in New Mexico and flows through Arizona, has been the subject of phreatophyte management and mismanagement since the early 1940s. By that time, tamarisk had completely overwhelmed the riparian ecology of the river and grew on several thousand hectares of floodplain and channel surfaces (Turner 1974). War-time demands for accelerated copper production prompted the search for "new" water supplies to be used in mining and milling in the Gila River Basin, and hydrologists suggested that phreatophyte removal might generate augmented supplies (Gatewood *et al.* 1950). Although little clearing was done during the war, post-war agricultural expansion kept the possibility of water salvage alive in state and federal governments.

In the 1960s the U.S. Geological Survey undertook a decade-long multi-million dollar project to assess the economic feasibility of water savings through phreatophyte clearing (reviewed by Culler 1970). Using the water balance approach in a study reach several km long on the Gila River immediately upstream from San Carlos Reservoir (behind Coolidge Dam), the survey attempted to measure evapotranspiration losses before and after a clearing project. With data from scores of stream gauges, rain gauges, and recording wells the study provided data for 414 two- to three-week periods, the most extensive data set ever collected for phreatophyte research.

The final report of the Geological Survey research (Culler *et al.* 1982) showed that in the study reach the phreatophyte cover, mostly tamarisk but with some willow and cottonwood, resulted in evapotranspiration of 81.2 cm (32 in) of moisture per year. In other words, the study area would be covered by a layer of water 81.2 cm deep if all the water lost through evapotranspiration were accounted for. After clearing, the water loss through evapotranspiration was reduced to 35.6 cm (14 in), implying a water "salvage" of 45.7 cm (18 in).

There are two reasons that dependable water savings are likely to be close to zero, however: variability of the measurements and the need for replacement vegetation that would utilize all the water savings, making use of the cleared land economically unprofitable. If sampling and measurement errors are taken into account and a 95% confidence interval applied, the figure for water savings ranges from -10.2 cm (-4 in) to +104.1 cm (+41 in), hardly useful input for planning and cost/benefit analysis.

U.S.G.S. workers reported that replacement vegetation was required because a barren floodplain and channel was an unnatural condition (Culler *et al.* 1982), and it is likely that phreatophytes would reinvade the cleared area if it were to be left unoccupied. They had no success in establishing replacement grasses, and ten years after the project, the cleared area was completely overgrown with phreatophytes. An economic analysis of crop and non-crop revegetation species by Patten (1984) showed that within any realistic scenario costs would exceed benefits.

The brief review of riparian vegetation provided in this chapter clearly indicates that ecological variability over space and time are unavoidable considerations in the environment-society interactions in the Colorado River Basin. Inadvertent alteration of the riparian community through the introduction of exotic species resulted in slow initial changes followed by rapid replacement of natural species by introduced plants. Attempts to return the riparian systems to former conditions appear to be either impossible or prohibitively expensive. It may be that the only reasonable course of action is no action at all, and that the most desirable condition is the continued growth of phreatophytes that we have spent 40 years trying to eradicate.

# 4

# Vertical Channel Instability: Arroyo Development

Environmental systems are intricately connected to each other so that changes in one system are propagated through many others. Instability in climatic, hydrologic, and riparian vegetation systems is strongly connected to instability in river channels, the subject of the next two chapters. The processes and responses among the various systems are bi-directional; that is, sometimes channel changes cause adjustments in other systems and sometimes changes in the other systems produce channel changes. The subjects of the following chapter are the processes and causes of vertical channel instability, the channel cutting and filling that produces arroyos throughout the Colorado River Basin.

## Mechanisms

An arroyo is a trench with steep sides, a rectangular cross section, and a stream channel along its floor. It is created by the erosion of water flowing in the channel and is excavated in valley-floor alluvium (Antevs 1952; Bryan 1922). Arroyos in the Colorado River Basin have depths as great as 21 m (70 ft) and lengths of several km. The term "arroyo" originally meant "stream bed" in Spanish but during the exploration of the southwestern United States in the nineteenth century it was used to denote the entrenched channels of some valley floors (Murray et al. 1933). Dodge (1902) introduced the term into the scientific literature beginning more than 80 years of debate on the "arroyo problem": what caused arroyos to develop and how do they operate?

The significance of arroyos for environmental management is that their development destroys valuable floodplain areas that represent the only useful farm lands in the arid Southwest. Irrigation from streams flowing near the surface of the floodplain is relatively easy, but if the stream becomes entrenched in an arroyo, the water flows several meters below the level of the fields on the surface of what was once a floodplain but which is now a terrace. Erosion of arroyos also mobilizes large quantities of sediment which fill reservoirs downstream and which seriously affect water quality.

There appear to be two basic types of arroyos: those that develop from discontinuous local erosion and those that are produced by base level lowering and system-wide headcut migration. Discontinuous arroyos develop when the local gradient along a stream becomes oversteepened by the deposits of sediment (Leopold et al. 1964). Erosion begins in this oversteep section and proceeds upstream, creating a trench with a more shallow gradient. When viewed in profile, the most remarkable feature of the system is the headcut, a place where on the long profile a waterfall occurs (Brush and Wolman 1957). The majority of erosion occurs in this nearly vertical section of the profile. As streamflow cascades over the headcut, the development of a plunge pool allows undercutting to occur, sediment is removed, and the headcut moves upstream. Sediment from this erosion process is transported a short distance downstream and is deposited. A number of these discontinuous trenches with eroding upper parts and depositing lower parts may eventually join each other to form a continuous trench with a new system-long gradient (Schumm and Hadley 1957). In other cases the headcut becomes progressively smaller and virtually disappears as it migrates up the stream into sections with less water for erosion.

Headcut erosion can also be initiated by a lowering of base level, which occurs when channel entrenchment along a master stream is caused by flood erosion or human manipulation. After the master stream is lowered, all the tributaries join it at discordant junctions that resemble headcuts. These headcuts then migrate upstream throughout the system in an erosion process as described above. Eventually long profiles of all the tributaries join the master stream at its level, but arroyos have been excavated throughout the system in order to accommodate the adjustment (see Strahler 1956 for discussion of profiles in gullies and arroyos; Shulits 1941, for the general case).

Because of the importance of the headcut migration process, management of the arroyo is usually concentrated on stabilizing the headcut and steep side walls of the arroyo (Heede 1974). Vegetation management is an important tool because the roots of trees and shrubs provide a useful stabilizing influence. If the natural riparian vegetation is altered by human activities or by climatic change, the resistance of the fluvial system to erosion is also altered. Hence, the interest of New Mexico conservationists in importing tamarisk in the early twentieth century as an erosion control measure is understandable (Robinson 1965). They hoped that the plant would stabilize arroyo banks and headcuts, a function that it performed at significant environmental cost as outlined in the previous chapter.

The Fremont River, a tributary of the Dirty Devil and Colorado rivers in southcentral Utah, provides an example of the processes and problems of arroyo development. In the early 1880s administrators of the Church of Jesus Christ of Latter Day Saints (Mormons) in Utah experienced population pressure as thousands of converts migrated to the state. The church decreed that the valley of the Fremont River should be colonized to provide a new settlement area and to serve as an agricultural production area. During the late 1880s the floodplain of the river provided easily irrigated land to which water was directed by simple brush dams across the channel (Snow 1977; Hunt *et al.* 1953). Tributary valleys had alluvial floors covered by grass for grazing the community herds.

In September 1896 a large flood on the Fremont River caused extensive erosion along the channel which excavated itself to a depth 7 m (22 ft) below its original level. In the next decade a number of floods caused the channel to increase its width by eroding the exposed arroyo banks. By 1910 the Fremont River had been converted from a 30 m (100 ft) channel flowing on the surface of a floodplain to a trench 7 m (22 ft) deep and 400 m (1320 ft) wide (Graf 1983b). Irrigation works were destroyed and fields as well as most of a townsite were eroded away.

Lowering of the bed of the Fremont River triggered headcut erosion of the tributary streams. By 1920 all the major tributaries had experienced at least two episodes of

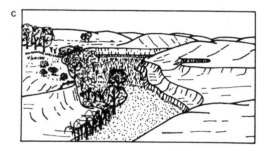

**FIGURE 8**     CHANGES IN THE CHANNEL OF PLEASANT CREEK, A TRIBUTARY OF THE FREMONT RIVER, SOUTHCENTRAL UTAH. **A:** In the late 1880s Pleasant Creek was a narrow meandering stream flow on top of alluvial fill. Redrawn from a sketch by C. B. Hunt based on eyewitness descriptions of early settlers and published in Hunt *et al.* (1953). **B:** In 1936 the creek had excavated and partially refilled a large arroyo. Redrawn from a sketch by C. B. Hunt based on his photograph of the site and published in Hunt *et al.* (1953). **C:** In 1982 the creek had experienced further infilling and renewed cutting. Drawn from a photograph by W. L. Graf of the same site shown in the previous views.

downcutting with little refilling (Figure 8). The result was that the tributary valleys were occupied by arroyos that incised their floors, impeding travel and lowering the water tables (which were coincidental with the elevation of the channel floors). With the lowering of the water tables, the grass disappeared and grazing prospects declined. Finally, the Mormon Church recalled its settlers and the valley was virtually depopulated.

## Temporal Structure

Although arroyos are frequently considered to be the product of unwise environmental management by modern society, they developed in the Colorado River Basin before the arrival of technologically advanced human cultures. Evidence of prehistoric arroyo cutting and filling has been preserved in sedimentary materials and structures in valley alluvium (Love 1979; Figure 9). In at least two areas of the basin the chronology of arroyo development since the beginning of the Holocene (the post Pleistocene period which began about 10,000 years ago) is reasonably well-known. In Chaco Canyon (Hall 1977), a New Mexico stream that is a southern tributary to the San Juan River, two complete cycles of cutting followed by filling occurred between 10,000 and 8,000 years BP (before present, with present considered to be 1950). About 8,000-7,000 BP another massive erosion episode occurred with arroyo development, followed during the period 6,700-6,000 BP by deposition and filling. Erosion of another arroyo in the canyon did not occur until about 2,400 BP, with refilling again about 2,200 BP. According to associated sedimentary evidence, increased precipitation and runoff was accompanied by arroyo cutting again about 850 BP, followed by deposition beginning about 600 BP. Finally, the well-known historical erosion episode began about 100 BP.

In San Pedro Valley, a southern Arizona tributary of the Gila River, a similar series of events is also documented by the valley sediments and their structure (American Quaternary Association 1976). From 10,000 to 8,000 BP deposition was common, with an erosion episode beginning about 8,000 BP. Deposition dominated from about 6,000 to 4,000 BP, and since 4,000 BP repeated cut and fill episodes occurred.

The fact that change in these two widely separated areas was at least broadly synchronous suggests that large-scale climatic changes were responsible for the arroyo processes. Researchers dealing with the problem of erosion and sedimentation have suggested that the erosion portions of the process resulted from increased runoff created by large amounts of moisture brought into the basin by zonal winds from the Pacific Ocean (Knox 1983; Hall 1977; Busby 1963). These circulation systems were apparently more common prior to about 8,000 BP, having occurred only occasionally since that time (Borchert 1950; Bryson et al. 1970). In more recent periods, meridional winds oriented in a north-south direction have dictated a more arid climate for the basin (Van Devender and Spaulding 1979).

The record of changes in arroyo processes over the last century is much more clear than the record for earlier times because of the availability of direct observations and in some cases photographic or map evidence (Gregory 1917, 1938, 1950; Gregory and Moore 1931; Hereford 1984). Data from these sources show that in 1880 arroyos were uncommon in the Colorado River Basin and that streams were relatively narrow, meandering channels flowing across the surfaces of valley fills. Between 1880 and 1900 channel erosion converted many of these streams into arroyos. Until about 1920 frequent floods were common in many areas of the basin, and minor amounts of fill would accumulate in arroyos only to be eroded again. From 1920 to 1940 the arroyos continued to conduct large amounts of sediment downstream, but their general configuration did not change. After the early 1940s, however, many of the arroyos began to fill with sediment. The refilling process appears to have been halted by a series of erosive floods that occurred in the late 1970s and 1980s, but it is too soon to

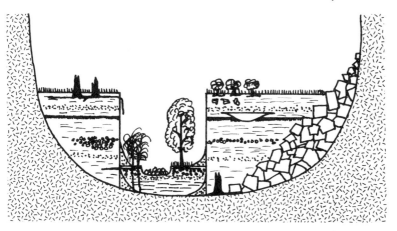

**FIGURE 9**   AN EXAMPLE OF SEDIMENTOLOGICAL EVIDENCE FOR PREHISTORIC ARROYO CUTTING AND FILLING. Sedimentary materials and structures exposed in an arroyo wall in Kane Spring Wash, a tributary of the Colorado River in eastern Utah.

determine whether this most recent change is significant. This record of the last century's changes in arroyo processes shows the unstable nature of the features and indicates that changes are abrupt and related primarily to floods.

The changes in arroyo forms are closely related to sediment yield from the drainage basins in which the arroyos occur because arroyos are created by the erosion of previously stored sedimentary materials. These materials are washed downstream as part of the sediment load carried by the river system (Colorado sediment processes reviewed by Howard 1947; Condit *et al.* 1978; Iorns *et al.* 1964, 1965; Oka 1962; Smith *et al.* 1960). Before the construction of high dams on the Colorado River, these sediments were ultimately carried to the delta of the river at the head of the Gulf of California. After the closure of Hoover Dam (1936), Glen Canyon Dam (1962), and Flaming Gorge Dam (1962), the sediments were trapped in reservoirs occupying space needed for water storage and are host to chemical pollutants. The management of the reservoirs therefore requires accurate knowledge of the sediment yield of the upstream basins.

In order to obtain the necessary information on sediment transport through the Colorado River system, the U.S. Geological Survey maintains one of the longest running sediment measurement programs in the world, but even it does not answer some of the managers' most important questions. Accurate prediction of future sediment yield depends on accurate characterization of past sediment yields so that statistical models can be constructed. The construction of the past sediment transport record is confounded by problems with measurement, changes in instrumentation, and the connection between arroyos and sediment yield.

Measurement of the sediment passing a given stream cross section is accomplished by lowering a trapping device into the water (Guy and Norman 1982). The device has a bottle which is filled with water and sediment. After removal from the stream, the sediment in the bottle is allowed to settle, it is removed, measured, and then compared to the total volume of water that accompanied it. This measurement is of sediment concentration per unit volume of water; it is multiplied by the total volume of water passing the measurement station to determine the total amount of sediment transport (Porterfield 1972). Inaccuracies occur in this method because the trapping device disrupts the flow of water which may cause too much or too little material to be collected. Of necessity, the method also ignores the materials being transported as bedload on the floor of the channel where particles may be transported by saltation (bouncing) or traction (rolling or creeping). Underprediction of the total sediment load results (Stelezer 1981).

A change in instrumentation also presents problems in interpreting the record of sediment transport in the Colorado River system. As shown in the time series in Figure 10 there appears to be a significant change in sediment yield in the early 1940s. Whether this change is a product of natural processes or of measurement technique is unclear, however, because in 1943 a new sampling device was introduced that may or may not have collection properties similar to the original. Tests have shown that the two devices are similar (St. Anthony Falls Hydraulic Laboratory 1957), but whether the experiments are applicable to field conditions in the Colorado River system is unclear.

Finally, the connection between arroyos and sediment yield makes the interpretation of the temporal changes difficult. The early 1940s is not only the time of change in instruments, it was also the time when filling began to occur in the arroyos upstream from the measurement stations (Graf 1983a). In addition, it was a time when riparian vegetation in the form of tamarisk was at its densest development in many areas, increasing the likelihood of sediment trapping and stability (Hadley 1961; Smith 1981). Sediment was beginning to be stored in the valley floors instead of being conducted downstream. The frequency of large floods apparently forms the connection between the two phenomena of sediment storage and vegetation change. Unlike humid regions where moderate flows (with recurrence intervals of 10 years or less) transport as much as 90% of the sediment, in arid and semi-arid regions like the Colorado River Basin, moderate flows account for only about 40% of the sediment transport (Neff 1967). Classical mathematical models are therefore difficult to apply to the Colorado River. These difficulties with the historical record make predictability of the temporal change in sediment yield for management and planning problematic at best.

## Spatial Structure

Throughout the Colorado River Basin, arroyo development takes on a particular spatial structure because the mechanisms that control it operate at a particular scale. The smallest streams in the basin (those with drainage areas of less than 1 sq km (0.4 sq mi) simply collect materials from the surrounding slopes and conduct them downstream without internal storage. These small basins respond mostly to local changes in vegetation and land management. Sectional streams, those confined to a small scale geologic-climatic section and with drainage areas of 1-1,000 sq km (0.4-386.0 sq mi), respond to local climatic changes and supplies of sediment from above. They are usually the locations of valley fills and arroyos.

Regional streams are those that drain complex geologic and climatic regions 1,000-10,000 sq km (386-3860 sq mi) in area. They respond primarily to flood events

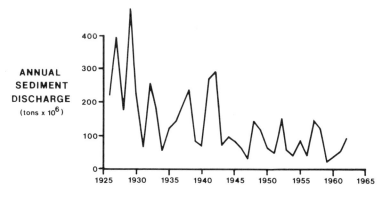

ANNUAL
SEDIMENT
DISCHARGE
(tons x 10⁶)

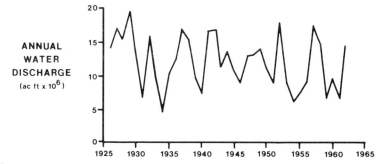

ANNUAL
WATER
DISCHARGE
(ac ft x 10⁶)

**FIGURE 10** TIME SERIES OF SEDIMENT TRANSPORT AND THE AN-NUAL FLOOD SERIES FOR THE COLORADO RIVER IN THE GRAND CANYON, ARIZONA. Data from U. S. Geological Survey Water-Supply Papers. The sediment record ends in 1962 with the closure of Glen Canyon Dam a short distance upstream. Note the association of high sediment discharges with high annual floods.

and also store sediment with occasional arroyo development. Finally, inter-regional streams with drainage areas greater than 10,000 sq km (3860 sq mi) respond only to alterations in the hydrology of their mountain source areas and to changes in the sub-continental scale atmospheric circulation. From the standpoint of the water resource user, changes in the smallest and largest scales are probably the easiest to predict, with the crucial medium-scale systems the most difficult.

Individual arroyos also have a definable systematic spatial structure. The depth of erosion is directly related to the amount of shear stress exerted on the channel floor or to the more easily measured surrogate of discharge. If the arroyo is the product of headcut migration its depth is greatest near where the headcut is initiated and declines as the headcut migrates upstream (Brush and Wolman 1957). Therefore depth of erosion is also inversely proportional to the distance from the point of initiation for the headcut (Graf 1982b).

## Causes

The exact cause of arroyo development is of interest to the land and water manager because of its destructive processes; furthermore, if the cause is known effective preventive measures may be possible. The cause of arroyo development is of interest to the environmental scientist because by understanding arroyo processes insight into more general environmental processes might be gained. The interconnections among climate, hydrology, vegetation, stream channel processes, and human activities are so complex, however, that determination of the cause of arroyo development is a difficult if not impossible task (review by Cooke and Reeves 1976). Many researchers investigating the question of causality have begun with a limited set of multiple working hypotheses that encompass the likely causal mechanisms: overgrazing, climatic changes, and internal adjustments inherent in the operation of channel systems in arid environments.

The first widely accepted explanation for the development of arroyos in the Colorado River Basin was that overgrazing of upland slopes caused increased amounts of runoff during storm events (Rich 1911; Leopold 1921). This water then rushed down hillslopes and concentrated in the valley-floor channels which were not adjusted to accommodate the large amounts. Enlargement of these channels by erosion was the logical outcome, aided by further overgrazing of the valley floors which reduced the cover of vegetation that bound soil masses and prevented erosion. The grazing hypothesis was attractive because some of the arroyo development occurred shortly after the introduction of large numbers of cattle into the basin and the destructive effects of improper stocking levels on the vegetation cover was obvious, even to the cattle owners (Atherton 1961). Along the Gila River drought forced intensive use of the river bank areas for grazing, and when floods ended the drought they also ended the previous channel configuration by extensive bank erosion (Swift 1926).

The grazing hypothesis cannot be fully accepted as the only explanation of causality, or even as a widely applicable one. In New Mexico and Arizona, for example, large numbers of cattle were introduced by Mexican herders in the late 1700s without extensive arroyo development (Denevan 1967). Conversely, erosion of arroyos had occurred before the introduction of cattle (Bryan 1941). Even when American cattle entered the range, not all areas experienced erosion. Finally, arroyo development has occurred in areas without grass cover which have never experienced grazing (Gregory 1950). The grazing hypothesis therefore works for some cases but not all.

A second hypothesis for the cause of arroyos called on climatic change. In this explanation as climate changes from relatively moist to relatively dry, the upland vegetation declines in density and runoff increases to erosive proportions as in the overgrazing hypothesis (Bryan 1922; Hack 1939; Leopold and Bull 1979). Although less rainfall occurs, more of that which falls runs off so the total runoff is greater than before. Other workers have reversed the scenario and suggested that as a change from relatively dry to relatively moist conditions occurs the newly copious rainfall produces large amounts of runoff until more dense vegetation establishes itself (Huntington 1914; Martin 1963). In each of these approaches, more runoff results in more channel erosion. Still other researchers have suggested that the intensity of the rainfall is the significant factor in arroyo development, and that as rainfall intensities increased, so did channel erosion (Leopold 1951; Cooke and Reeves 1976).

Climatic hypotheses are appealing because palynological evidence clearly shows that climate has changed in the Colorado River Basin on a variety of time scales. Furthermore, climate cannot be ignored since it provides the basic mass and energy inputs of the fluvial systems involved. The hypotheses are not intellectually satisfying, however, because climatic changes are generally considered to occur over broad areas synchronously while arroyo initiation occurred at different times throughout the basin (Bryan 1922). There have not been enough shifts from moist to dry periods and back again to provide adequate data to test the expected effects (Heede 1976). Finally, changes in rainfall intensity have been relatively minor during the instrumented record and do not seem to be statistically signficant (Knox 1978).

The third major hypothesis for arroyo development suggests that alternative cutting and filling are part of the normal processes found in arid and semi-arid watersheds and that they would occur without any adjustments in climate or grazing (Schumm 1973). Early workers assumed that stream systems operated on a cyclic basis, while modern workers have used laboratory simulations with physical models to show that highly variable rates of sediment production (and thus presumably erosion) are common even in undisturbed steam systems (R. A. Parker, cited by Schumm 1976).

As with the other hypotheses, the internal adjustment concepts have limitations. The laboratory experiments have addressed only very small systems, so that the relative importance of internal versus external influences and variations due to scale differences cannot be evaluated (Church and Mark 1980). Natural systems might require a trigger mechanism such as a climatic change or a large flood to initiate the adjustment, so that explanation of the timing of the arroyo development depends on explaining the trigger mechanisms.

This brief review has not touched on all the proposed explanations, but recent workers have concluded that arroyo initiation can be caused by all of these mechanisms or by any combination of them (Cooke and Reeves 1976; Heede 1976; Graf 1983b). Large floods almost always play a role, but beyond that, no broad generalization is possible. Also, each of the causal mechanisms produces the same end product, an arroyo, so the arroyo process is termed one of "equifinality" (Cooke and Reeves 1976). The original cause cannot therefore be determined by simple analysis of the result. Overgrazing can be prevented through wise management, but it is not likely that environmental managers will be able to control climatic change or inherent internal adjustments in stream systems. Throughout the Colorado River Basin the management solution to arroyo development is limited to restricting as much as possible the extent of the erosion and to avoiding those hazardous locations where catastrophic erosion may occur (Heede 1976).

Whatever their cause, arroyos have far-reaching human implications. Their destruction of valuable lands and their role in the processes of sediment transport and storage make them the subject of environmental management concern. Like most environmental systems in arid and semi-arid regions, arroyo and sedimentation processes are characterized by long periods of quiescence broken by catastrophic changes that are difficult to predict. In the Colorado River Basin catastrophic instabilities in the climatic, hydrologic, riparian vegetation, and fluvial systems are all intimately related through the connective processes associated with water.

# 5

# Horizontal Channel Instability: Channel Migration

Unstable environmental conditions that caused arroyo cutting during the past half century in small and medium sized streams in the Colorado River Basin also caused horizontal channel instability in large steams by initiating channel migration across floodplains and channel widening through bank erosion. These destructive processes resulted in the loss of valuable agricultural, grazing, and townsite lands in the early twentieth century, and continue to pose hazards in urban and rural settings. The following chapter begins with a review of the mechanisms of horizontal channel instability, continues with a discussion of floodplains in the arid and semi-arid environment, and concludes with an accounting of the special problems that such instability presents in modern urban areas.

## Mechanisms

Two types of channel form commonly encountered in arid and semi-arid drainage basins are meandering and braided (Leopold and Wolman 1957; Lane 1957). Meandering channels are usually relatively narrow and shallow and were characteristic of many of the late nineteenth century streams in the Colorado River Basin (Dutton 1882). Over-bank flows were common in such streams, with sediment deposited along the channel sides as aggrading floodplains. In the early twentieth century braided channels became more common (Gregory 1950). These relatively wide versions with various depths resulted from significant erosion and transportation of sediments. In more recent times, transitional forms between the two extremes have been most common (Figure 11).

The changes from meandering to braided channels that produced catastrophic erosion on near-channel lands represent stream system adjustments in response to several controlling factors (Schumm 1977). The timing, frequency, and magnitude of floods play a major role supplemented by alterations in sediment supply from surrounding slopes, channel gradient adjustments, and changes in riparian vegetation. Floods are important because they have the large amounts of energy needed to cause rapid erosional changes. If water overflows the banks of a small meandering channel and does not erode large amounts of bank material because that material is effectively anchored by vegetation, few channel changes occur. On the other hand, if surface protection is not available in the form of dense vegetation, large floods cause destruction of the original meandering channel with a wider, more straight, braided version excavated by erosion.

Fremont River at Giles, Utah (T28S, R10E, Sec. 21)

1883

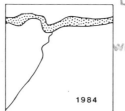

1984

San Juan River at Bluff, Utah (T40S, R21E, Sec. 35)

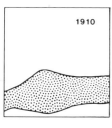

1910

1984

Gila River near Gila Bend, Arizona (T2S, R5W, Sec. 21)

1887

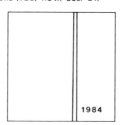

1984

**FIGURE 11**   EXAMPLES OF CHANNEL TYPES AND CHANGES IN THE COLORADO RIVER BASIN. The Fremont River changed from a suspended load to a bedload stream as a result of a catastrophic flood in 1896 (1883 data from General Land Office Surveys; 1984 data from field observations). The San Juan River shows a change in the opposite direction resulting from sedimentation and invasion by phreatophytes (1910 data from photos by H.E. Gregory; 1984 data from field observations). The Gila River has been altered by irrigation withdrawals and channelized to control its channel location (1887 data from General Land Office Surveys; 1984 data from field observations).

If no large floods occur for several years or decades, the braided channel may be slowly filled by sediments deposited there by low discharges (Schumm and Lichty 1963). If there are no interruptions by recurring large floods, the braided channels may slowly be converted once again to narrow, shallow meandering ones. Channel restriction may also be accelerated by constant withdrawal of waters in upstream areas for irrigation purposes, so that average discharges under post-settlement conditions are much less than in pre-settlement periods. The sequence of events outlined above appears to be fairly typical throughout the Colorado River Basin.

Channel gradient changes may also be reflected in channel forms (Schumm et al. 1972). When large amounts of sediment from surrounding slopes are washed into the upper reaches of a river, they may accumulate there, causing the long profile of the stream to adopt a steeper configuration. This arrangement produces more rapid flows and attending greater erosive capability which results in changes from meandering to braided forms. Such infusions of sediment in the Colorado River Basin have frequently resulted from overgrazing or climatic changes which have reduced protective vegetation cover on slopes and thus released large amounts of sediment (Lusby et al. 1971). Mining and milling activities also have produced great heaps of easily eroded sediments (Nelson and Shepherd 1982).

Riparian vegetation changes as described in Chapter 3 also affect the mechanical stability of river form (Smith 1981). Because many channels in the Colorado River Basin are established in unconsolidated materials made of alluvial fill, variation in bank stability is frequently related to the nature and density of vegetation growing there (as well as to sediment particle size). In the Lower Basin mesquite forests commonly lined channels adding stability to their perimeters, while in the middle basin, cottonwood forests were common and in the Upper Basin, at the highest elevations, willow was found (Hastings and Turner 1965). With the advent of Anglo settlement, the trees were cut for lumber and fence posts, and cattle grazed on the dense grass that once grew under the trees. Large floods that had produced little or no channel changes with the vegetation in place proved to be destructive when they occurred in the absence of dense riparian vegetation. When exotic vegetation, especially tamarisk, invaded the basin, it replaced much of the vegetation that had been artificially removed and introduced a stabilizing influence. Braided channels became less common, and a transition back to meandering ones became apparent.

Simple cause and effect connections among interacting system elements are relatively difficult to establish with certainty. The case of the Salt and Gila rivers in central Arizona is a typical example (Figure 12; Graf 1982a). In the middle 1800s prior to Anglo settlement the stream in the vicinity of Phoenix had a highly variable discharge because it derived much of its waters from snowmelt in the mountains to the east. Even in summer, however, it usually carried about 22 cms (800 cfs), enough to necessitate the development of ferry crossings for the well-defined channel (Powell 1893). After about 1870 woodcutting and grazing reduced the riparian vegetation cover, and in 1891 a flood of 8,400 cms (300,000 cfs) created a braided channel nearly 1.6 km (1 mi) wide.

The braided channel was maintained for almost four decades by frequent floods of large magnitude, but by the late 1930s the channel began to narrow again as tamarisk invaded the area, stabilized the banks, and trapped sediment in the channel. In 1938 the last of six high dams upsteam was closed, and in the summer months the channel had no water. Relatively narrow, well-defined channels resulted, and the braided form was less in evidence. From 1941 to 1965 there was no sustained flow in any month, groundwater tables declined under heavy pumping without recharge (Aldridge 1970), and the riparian vegetation slowly disappeared (Graf 1983b). Lulled into a false sense of security, urban and agricultural developers moved their operations into the near-channel areas. Minor floods in 1965 and 1973 resulted in little more than apprehension, but several floods (one of 5,040 cms; 180,000 cfs) in the late 1970s and early 1980s

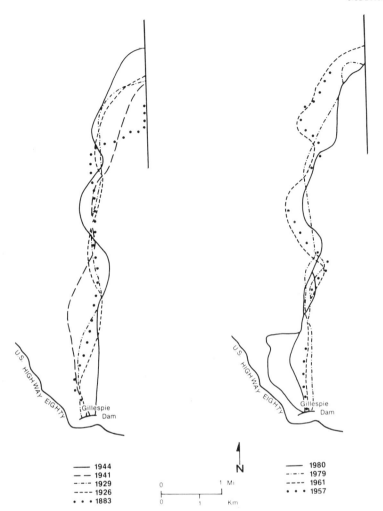

**FIGURE 12**    CHANGES IN THE CHANNEL LOCATION OF THE GILA RIVER, ARIZONA, 10 KM ABOVE THE TOWN OF GILA BEND, ARIZONA. Note radical changes in the northern reaches between 1883 and 1926 (probably caused by erosion during floods) and in the southern reaches between 1979 and 1980 (probably caused by sedimentation during floods). Data from General Land Office Surveys, irrigation surveys, and aerial photographs.

caused damage of over $100 million (Aldridge and Eychaner 1984; U.S. Army 1979a, 1979b, 1980). In addition, the familiar braided channel was re-established. River planners and managers must again accommodate a channel which is a kilometer and a half wide.

## Floodplains

Because floodplains are common features of the meandering rivers in Europe and the eastern and midwestern United States, and because most of the body of laws governing the United States was derived from those regions, the concept of floodplain is firmly embedded in the institutional structures attempting to manage the Colorado River Basin (general review of floodplains by Lewin 1978). The concept is not useful when dealing with braided channels, arroyos, and systems that are highly changeable, however. The clash between human expectations and actual processes regarding floodplains in the Colorado River Basin has not been a successful merger of society and environment, and therefore bears some examination.

There are three commonly recognized types of definition for floodplains: geometric, hydrologic, and geomorphic (Ritter 1978). Geometrically, a floodplain is any relatively flat surface next to a channel and separated from the channel by a bank. The advantage of this definition is that floodplains can be easily recognized in the field, but it includes rock surfaces as well as valley fill materials and does not recognize the frequency of inundation. Hydrologically, a floodplain is a surface that is inundated by water from a channel within a given recurrence interval. The ten-year floodplain is therefore covered by waters on average once every ten years, or stated differently, such surfaces have a ten percent probability of being inundated each year. The hydrologic definition is useful because it has significance for planning and management efforts, but it does not indicate the stability of the surface involved. Finally, the geomorphic floodplain is a relatively flat surface next to a stream channel, separated from it by a bank, and built of materials transported and deposited by the stream in its present regime. The geomorphic definition is useful because it indicates the parts of the landscape that are potentially unstable, but it is expensive to apply because detailed mapping of soils is required.

Each of these definitions was applicable in humid regions or in the semi-arid Colorado River Basin before the recent channel instability occurred. Once arroyos were excavated and channel widening and migration became features of the landscape, however, the term floodplain and its convenient conceptual attachments became problematic. In the case of an arroyo, what was once the surface of the floodplain had become a terrace (Happ 1971) that even the 500-year flood would not inundate. In larger streams the braided channel frequently occupies the entire valley floor, and the banks of the channel are coincidental with the edges of terraces or hillsides (Miall 1977; Miller 1970). In each of these cases floodplains of the traditional variety do not exist (see review by Lattman 1960), yet planners and managers attempt to apply the concept for lack of any reasonable substitutes.

The radical nature of the changes in Colorado River Basin streams and their floodplains are exemplified by two examples: the San Juan and Paria rivers. The San Juan River flows westward from the San Juan Mountains of Colorado to join the Colorado River in southeastern Utah. In 1879 it was the objective of a special settlement effort by pioneers from western Utah, in part because of the anticipated stability and productivity of its floodplain (Miller 1966). The excavation of arroyos in tributary streams dumped huge quantities of sediment into the main stem of the San Juan River, however, and frequent floods caused its channel to widen to several times its presettlement width through intensive bank erosion (Miser 1924a, 1924b). The channel became a vast ribbon of sand occasionally filled with water and encroaching on the already limited fields. Portions of the town of Bluff, Utah, literally were washed away along with most of the irrigation works. Eventually most of the settlers moved to other more promising locations, leaving the valley floor to the invading tamarisk.

The town of Paria was located on the Paria River, a stream that begins near what is now Bryce Canyon National Park and that joins the Colorado River at the head of the Grand Canyon (following account from Gregory and Moore 1931; Woodbury 1944). Established in 1871, Paria was an important outpost linking Utah and Arizona with a bustling population of as many as 500 people in its town and associated river-valley farms fed by irrigation waters from the narrow meandering channel. In 1883 floods began the excavation of the channel, creating an arroyo about 5 m (15 ft) deep which then widened by lateral erosion. By the early 1900s the entire town had been washed away and all but a few hectares of farmland had been destroyed. In 1984 one farm house remained in the valley that had become almost entirely river channel. If changes of this magnitude have occurred in the past century, equal changes are possible in the next.

## Floodplain Management

The concepts surrounding the term floodplain enter into environmental management of the Colorado River Basin in a legal sense. In legal approaches, state and federal laws designate certain areas near the channel as "floodplain" and impose restrictions on the use of such areas (Office of Emergency Preparedness 1972). The National Floodplain Insurance Program provides some financial protection for users of floodplains but only if certain precautions are taken. Buildings must use special flood-proofing construction, for example (Sheaffer 1967). Users of floodplains who do not observe the restrictions cannot purchase the insurance and are not likely to obtain affordable substitute coverage from private companies who are aware of the substantial risks involved. For these reasons, the designation of particular parts of the landscape as floodplain carries with it a considerable amount of legal baggage (White 1964).

In many areas the provisions of floodplain insurance programs are ignored, and land users erect buildings in hazardous locations without structural or financial protection (Kates 1962; Bauman and Elmer 1976). When flood damages occur, they then request disaster assistance from the federal and state governments. This arrangement was common in the 1983 floods along the lower Colorado River where many damaged buildings had been erected in hazardous locations without insurance. The risk-takers therefore were subsidized by the general society even though the benefits were clearly limited to only a few individuals. Disaster aid was limited by several federal and state agencies in recognition of the problem, but obtaining disaster assistance in the Colorado River situation soon became more of a political question than a financial or ethical one (Smith and Tobin 1979).

The legal definition of which parts of the landscape are to be administered as "floodplain" depends on the application of the hydrologic definition of the term and the use of mathematical models. Usually the 100-year floodplain, the surface near the channel that is subject to inundation at least once in every century, is designated as the restricted area. In the Colorado River Basin the usual approach is to use aerial photography to map the channel and near-channel areas in question in great detail with contour intervals of about 0.6 m (2 ft). Cross sections are generated from the maps or from photogrammetric techniques and are digitized. Hydrologic records produce an estimate of the amount of water expected in the 100-year flood, and that quantity is then used to mathematically fill each cross section to the 100-year flood elevation. The commonly used computer program, HEC2 (HEC = Hydraulic Engineering Center), indicates the elevations of the water surface which can be combined with topographic maps to outline the floodplain areas used by administrators and planners (U.S. Army 1981).

Difficulties with this approach derive largely from the fact that the technique and the HEC2 program were developed for humid-region streams and they are not necessarily applicable in semi-arid settings. In humid areas each stream has a single pair of banks to separate the channel from the floodplain, but in the semi-arid regions the braided channels have several banks and several channels, and sometimes no floodplain at all in the geomorphologic sense. In humid-region streams, water flows continuously and channel forms change very slowly; in the semi-arid situation, however, channels reflect mostly the influence of the last flow, so the mapped channel configuration is not likely to exist by the time the analysis is complete. Arid-region channels have riparian vegetation that strongly influences the characteristics of flow but that is not directly accounted for in the mathematical models. Finally, the hydrologic records in the Colorado River Basin are sometimes ineffective predictors of future discharges because the management of upstream dams may cause the unexpected release of large discharges.

The Agua Fria River, a south-flowing tributary of the Gila River in central-western Arizona illustrates the legal problems associated with floodplains on rivers in semi-arid regions (Wildan Associates 1981). As the river issues from the Bradshaw Mountains its waters are stored for irrigation purposes by Waddell Dam. The channel below the dam contained water for several months each year before the dam was closed in 1927. The braided, sandy channel was occupied by flows only once a decade on average thereafter and it appeared to be safe to some land users who farmed areas of the valley floor and built homes there. In 1976 an engineering company photogrammetrically surveyed the valley floor and produced for local governments a map of the 100-year floodplain for planning purposes. Floods in 1978 destroyed some property and when floods again occurred in 1980 the area of inundation bore little resemblance to the expected flood areas based on the original maps. Property owners sued the operators of Waddell Dam for being negligent in releasing inappropriate amounts of water; the longest civil court case in Arizona history resulted (data in following paragraphs from unpublished material in the case *Vittori et al. vs Maricopa County Water Conservation District Number 1*).

The maps of floodplain areas were not useful in the Agua Fria situation because the mathematical program and the concepts behind it were designed for humid-region conditions. The channel of the Agua Fria River is almost a kilometer wide in some places yet only a few meters deep. It has numerous islands, bars, subchannels, and small banks, and its only real boundaries are the edges of the Pleistocene terraces on either side. In the humid-region sense of the word, the river has no floodplain, only a channel. Land users who constructed buildings on the valley floor actually built them in the channel, so that when water releases from the dam occurred in 1978 they sustained damages.

The flows in 1978 also changed the channel configuration by eroding some parts of the channel and depositing materials in other parts. The topography upon which the floodplain maps were based was completely changed, so that when flows occurred in

1980 the system behaved entirely differently than expected, and accurate predictions based on the old maps were impossible. The importation of concepts and technologies from the eastern and midwestern United States provided little understanding for the management of a western landscape.

## Migrating Channels in Urban Areas

The natural variability in system behavior for semi-arid streams is enhanced in urban areas where human manipulations of the channel introduce further changes. Channelization, bridge maintenance, and sand and gravel mines are major contributors to human-induced instability. Channelization consists of hardening the boundaries of the channel, usually with cement or riprap (large blocks of cement or stone laid along the channel edges), and includes realignment. The system instability that results from these efforts sometimes includes the destruction of the engineering works designed to accomplish the channelization (Yang and Song 1979; Keller 1976). The straightened, narrowed channels have deeper, faster flows than their natural predecessors, permitting accelerated erosion of the channel form. Most rivers have some naturally occurring meanders that allow the system to adjust its gradient to the amount of water and sediment moving through the channel, so the imposition of straight channels represents an unstable condition. Equilibrium is usually restored by the re-establishment of some minor meanders through erosion of channel banks which also causes destruction of property.

Bridges pose special problems in semi-arid regions because of the discontinuous nature of flow. Citizens accustomed to dry channels are reluctant to tax themselves for large bridges that are likely to be needed less than one percent of the time. Frequently this results in the construction of underbuilt bridges — structures that are too small for the flows that occur with modest frequency (*e.g.*, 10-year flows). The flow is forced through openings under the bridge structures that are too small to accommodate the water, and the erosion that results ultimately may undermine the entire structure. Even if the structure is large enough, it may have to be set in unconsolidated valley fill instead of anchored on bedrock. When floods occur, the materials below the river bed are mobilized to a depth twice that of the water flow, so that bridge piers become unstable and threaten the stability of the structure.

Sand and gravel mines are common in river beds of the ephemeral streams in the Colorado River Basin because of the utility of these materials for construction. Clean sand and gravel can be taken directly from the beds of the rivers when they are not flowing, and since most urban areas are located along river courses, transportation costs are reduced. The large pits left behind after mining operations are completely refilled during each flood, providing the owner with a renewable resource. Unfortunately, the pits also drastically alter the configuration of the channel (Bull and Scott 1974), replacing smooth planar surfaces with gaps 10 m (30 ft) deep that can redirect flood flows away from natural courses, cause extensive erosion, and make floodplain predictions useless.

Flooding problems in the semi-arid Colorado River Basin demand a re-orientation of perspective by planners and managers away from experiences in humid regions where flood damage is mostly related to inundation. In the dry areas inundation is only a limited part of the problem. Channel erosion in the horizontal and vertical dimensions requires at least equal and probably more consideration in predicting and assessing potential flood damages.

Rillito Creek, a tributary of the Santa Cruz and Gila system that flows across the northern portion of Tucson, shows the special problems associated with semi-arid urban streams (Pearthree 1982). Urban development along Rillito Creek during the 1970s and 1980s focused local attention on the instability of the stream. In the late 1800s it had been a narrow, meandering, marshy stream lined with hay fields, but headcut migration upstream from a newly formed arroyo along the Santa Cruz resulted in the excavation of an arroyo along the Rillito several meters deep by 1910 (Cooke and Reeves 1976). Horizontal instability through lateral bank erosion has continued to the present.

Floodplain mapping along the Rillito by the U.S. Army Corps of Engineers and local agencies showed that inundation in some parts of the stream from the 100-year flood were extremely unlikely because the arroyo was large enough to contain all the waters from such a flood. When a flood of nearly the 100-year magnitude occurred in 1983 there were severe structural losses and damages resulting from erosion, however. Many of the structures damaged were not in the 100-year floodplain, but because erosion hazard zones had not been defined they were constructed in locations likely to sustain damage. Planning and insurance approaches to near-channel land management account only for inundation losses, but simulations based on past channel location changes indicate that in the next fifty years losses from erosion are likely to be five times greater than the losses from inundation along the Rillito (Graf 1984).

Horizontal channel instability is a feature common to streams in the Colorado River Basin that is not yet adequately accounted for in planning and management processes. Scientific analysis of the phenomena provides reasonable explanations, but engineering applications do not provide adequate predictions of river behavior that can be incorporated into political and financial processes of society. Until the naturally occurring variability can be included in societal accommodations, ephemeral rivers will continue to be viewed more as hazards than as resources.

# 6

# River System Instability: Impacts of High Dams

The Colorado River Basin contains numerous large dams designed to control the flow of the river system to enhance economic development. The construction and maintenance of these structures has brought about significant reductions in the variability of the flow of the river, but at the cost of increased instability in other aspects of the system. The following chapter briefly reviews the historical, political, and economic geography of high dams in the Colorado River Basin and then explores the impacts of the structures in upstream and downstream environments.

## Dams of the Colorado System

Two types of dams have been built on the Colorado River system: relatively low structures that do not create reservoirs and that serve only as diversion works to direct river flows into irrigation canals, and relatively high sructures that create reservoirs and significantly alter the flow of water in the channel (Burgess and Quirk 1978). Diversion works, ranging in size from small brush affairs to concrete structures up to about 10 m (33 ft) high, are common throughout the basin but their impacts are highly localized and so they are not the subject of further discussion here. High dams, on the other hand, represent some of the American society's largest engineering works and include massive concrete structures over 210 m (700 ft) high (Berkman and Viscusi 1973). Though not as numerous as diversion works, high dams have dramatic impacts on their riverine surroundings because they influence river flow and store huge amounts of water in their reservoirs (Table 3). Their construction also alters regional economics and creates the need for major transportation developments (Sutton 1968).

Justification for the construction of high dams in the Colorado River Basin has rested largely on the need for flood control, electrical power generation, irrigation, and to a much lesser degree, recreation (Bureau of Reclamation 1946; Robinson 1979). Valuable agricultural lands in southern California's Imperial Valley were continuously threatened by flood hazard in the early 1900s (Waters 1946). The completion of the Alamo Canal in 1900 insured that water would be conducted from the river to the irrigated valley, but the maintenance of the canal was complicated by high spring discharges (Freeman 1923; Howe and Hall 1910). When floods destroyed the designed opening of the canal in 1905 and the entire river was diverted into the valley it required two years, the political force of President Roosevelt, and the wealth of the

**TABLE 3**   MAJOR HIGH DAMS IN THE COLORADO SYSTEM

| River | State | Dam | Reservoir | Completed | Height (m) | Capacity (ac ft) |
|---|---|---|---|---|---|---|
| Colorado | AZ/CA | Parker | L. Havasu | 1938 | 98 | 655,000 |
| Colorado | AZ/NV | Davis | L. Mohave | 1950 | 61 | 1,839,000 |
| Colorado | AZ/NV | Hoover | L. Mead | 1936 | 221 | 30,092,000 |
| Colorado | AZ | Glen Canyon | L. Powell | 1964 | 216 | 27,306,000 |
| Colorado | CO | Granby | L. Granby | 1950 | 91 | 546,000 |
| Los Pinos | CO | Vallecito | Pine L. | 1941 | 49 | 131,000 |
| Blue | CO | Green Mountain | Green Mt. L. | 1943 | 94 | 126,000 |
| Agua Fria | AZ | Waddell | L. Pleasant | 1927 | 78 | 165,000 |
| Gila | AZ | Coolidge | San Carlos L. | 1928 | 76 | 1,222,000 |
| Verde | AZ | Horseshoe | Horseshoe L. | 1949 | 43 | 141,000 |
| Verde | AZ | Bartlett | Bartlett L. | 1939 | 87 | 182,000 |
| Salt | AZ | Stewart Mtn. | Saguaro L. | 1930 | 63 | 71,000 |
| Salt | AZ | Mormon Flat | Canyon L. | 1938 | 68 | 59,000 |
| Salt | AZ | Horse Mesa | Apache L. | 1927 | 93 | 248,000 |
| Salt | AZ | Roosevelt | Roosevelt L. | 1911 | 85 | 1,398,000 |
| Green | WY | Flaming Gorge | Flaming Gorge | 1964 | 153 | 3,832,000 |
| Strawberry | UT | Strawberry | Strawberry L. | 1913 | 22 | 232,000 |
| Price | UT | Scofield | Scofield L. | 1946 | 38 | 560,000 |
| San Juan | NM | Navajo | Navajo L. | 1963 | 123 | 1,728,000 |
| **Total Storage** | | | | | | 70,334,000 |

Data from International Commission on Large Dams (1973); includes reservoirs with capacity greater than 50,000 acre feet. Capacities differ from reports by Mermel (1958) and Bureau of Reclamation (1946).

Southern Pacific Railroad to re-establish control of the canal and river. Boulder Dam (later renamed Hoover Dam) was viewed by its builders as primarily a flood-control effort to prevent such catastrophes (Lingenfelter 1978).

High dams create the possibility for the efficient generation of electrical power because they create large vertical drops for the water which can be used to turn turbines for electrical generators (LaRue 1916, 1925). Hoover Dam was equipped with generators to produce electrical power for sale as a way of producing revenue to pay the cost of the dam. As electrical demand grew in the southwestern United States, the electrical power production by dams became an increasingly important part of their justification (Robinson 1979; Warne 1973). The construction of dams is therefore connected to the construction of electrical generating stations using fossil or nuclear fuels.

The latter point is demonstrated by the dams planned for the Grand Canyon (following account from Nash 1967; Fradkin 1981; Wyant 1982; Allin 1982). Electricity required to pump water in the Central Arizona Project from the Colorado River to Phoenix and Tucson was originally planned to have been generated by a series of dams in the Grand Canyon. These structures, called "cash register dams" because they would be relatively inexpensive to build and would generate large amounts of income from the sale of electricity, were the subject of intense political debate during the late 1950s and 1960s. The result was an agreement between federal agencies desiring to build the dams and environmental groups who waged a national campaign against their construction in Grand Canyon National Park. Both sides agreed that the dams would not be built (at least for the time being), but that the required electricity could be generated by coal-fired stations, including the Navajo Generating Station to be built near Glen Canyon Dam at Page, Arizona. The station therefore had the blessing of the preservationist lobby despite negative environmental impacts.

Flood control and electrical power production are directly related to the third major benefit of high dams: irrigation. By controlling the flow of rivers, the dams reduce the spring flood peaks which destroy irrigation works and store the water for use in summer and early fall when it is most needed for growing crops. The electricity generated by the dams is in part used for pumping irrigation water either to move it through surface systems or to draw additional water from subsurface sources (Johnson 1977).

Finally, a by-product of the establishment of reservoirs is the creation of the potential for recreation. Until the 1950s this benefit received little attention, but with the post-World War II economic boom in the West, recreation demands increased significantly and the promise of fishing, boating, and the creation of new wildlife habitats became a standard feature of justifications for high dams (U. S. Water Resources Council 1977a, 1977b).

The legal history of high dams in the Colorado River Basin is a tangled story of interstate rivalries and national investment in regional development. The Reclamation Act of 1902 resulted in the construction of the first high dam in the basin on the Salt River in central Arizona; Roosevelt Dam was closed in 1910 (Meredith 1968). By 1938 six more high dams were completed on the Salt and Gila system (Smith 1972). Dams on the Colorado River required interstate agreements. The Colorado River Compact of 1922 represented a compromise between Lower Basin states (primarily California) that wanted a high dam for flood control and Upper Basin states that feared rapid development to the south would result in establishment of claims to all the river's waters under the doctrine of prior appropriation (Jacoby *et al.* 1976). The Compact divided the river's water between the two halves of the basin. In recognition of this division of the resources, Congress approved the Boulder Canyon Project Act in 1928, which resulted in the completed Hoover Dam in 1936. The Metropolitan Water District of Southern California completed the Colorado River Aqueduct, and in 1939 the All-American Canal (with its route completely within the borders of the United States) replaced the Alamo Canal in conducting water to the Imperial Valley.

The Colorado River Compact did not apportion waters among the various states in the Upper and Lower Basins, so in 1948 the Upper Colorado River Basin Compact was established to distribute the Upper Basin share among five states (Table 4: following material review by Hundley 1964, 1966; Nadeau 1974). As with the Lower Basin states, Congress recognized the settlement by authorizing the Upper Colorado River Storage Project Act in 1956; this resulted in the completion of high dams at Curecanti, Flaming Gorge, Navajo, and Glen Canyon. Arizona and California could not agree on the disposition of the Lower Basin's waters, but after more than 30 years of congressional infighting and law suits (highlighted by calls to the Arizona state militia to defend the Arizona banks of the Colorado River) the U. S. Supreme Court decreed an apportionment in the case *Arizona vs. California* (Johnson 1977; Terrell 1965). Settlement of this issue resulted in the authorization of the Central Arizona Project Canal in the Colorado River Basin Project Act of 1968. Since that time, the Central California, Central Nevada, and Central Utah projects have been approved to distribute the river's waters to local areas.

## TABLE 4   DIVISION OF COLORADO RIVER BASIN WATERS

| State | Legal Entitlements (million ac ft per yr) | Actual Distribution[e] (million ac ft per yr) |
|---|---|---|
| California | 4.400[a] | 4.400[f] |
| Arizona | 3.800[a] | 2.050[f] |
| Nevada | 0.300[a] | 0.300 |
| **Lower Basin** | **8.500**[b] | **6.750** |
| Colorado | 3.881[c] | 2.406 |
| Utah | 1.725[c] | 1.070 |
| Wyoming | 1.050[c] | 0.651 |
| New Mexico | 0.844[c] | 0.523 |
| **Upper Basin** | **7.500**[b] | **4.650** |
| Mexico | 1.500[d] | 1.500 |
| **Total** | **17.500** | **14.500** |

Notes: [a]1928 Boulder Canyon Project Act
[b]1922 Colorado River Compact
[c]1948 Upper Colorado River Basin Compact
[d]1944 Mexico-U.S. Treaty
[e]Includes losses to evaporation of 0.6 million ac ft per year in the Upper Basin and 0.9 million ac ft per year in the Lower Basin, and 0.9 million ac ft per year inflow to Lower Basin from local streams
[f]Agreement at time of Central Arizona Project authorization by Congress. Upper Basin amounts agreed to by states as percentages.

Once identified, the specific locations of dams became a topic of intense political debate between the lobbyists for development and environmental groups (Nash 1967). In the Colorado River Basin high dams can be most efficiently constructed in narrow canyons where the entire river can be blocked at a constriction. Such locations usually coincide with reaches of steep gradient, rapids, and scenic beauty (Figure 13: Herron 1917). With the exception of Hoover Dam, all the high dams in the Colorado system were proposed during the 1950s and 1960s when the environmental movement was gaining popular support and refined political acumen.

Congressional hearings became battlegrounds between those concerned with the economic development promised by construction of high dams and those concerned with prevention of damage and loss of unique environmental resources (Nash 1967). Environmental organizations, especially the Sierra Club, Wilderness Society, and Audubon Society, were successful in preventing the construction of dams planned for the Grand Canyon and for the canyons of the Green River in Dinosaur National Monument (Nash 1970; Stegner 1955), but western congressional representatives won the dispute over Glen Canyon Dam which was completed in 1962 (Porter 1962). By the 1980s the Colorado River system had become the most regulated river network of its size in the world, but the upstream and downstream impacts of the control structures were just beginning to become apparent.

## Upstream Impacts

Dams directly affect the capability of river systems to transport sediment. The structures raise the local base level to create pools with still waters, causing deposition of materials previously transported by the river (review by Simons and Senturk 1976).

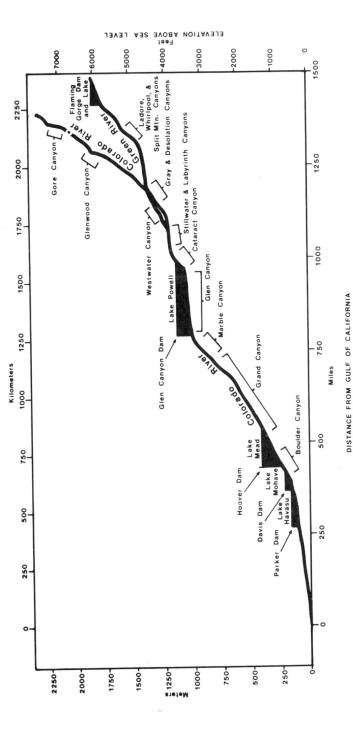

**FIGURE 13**   PROFILES OF THE GREEN AND COLORADO RIVERS SHOWING LOCATIONS OF DAMS, RESER-VOIRS, AND WHITEWATER CANYONS. Redrawn from LaRue (1916) and Bureau of Reclamation (1946).

More than 90% of the materials are deposited on the floor of the reservoir and rema
there throughout the life of the dam, slowly accumulating and filling the space original
intended for the storage of water (Brown 1944; Brune 1953; Churchill 1948; ar
Borland 1971 review trap efficiency of the process). Original estimates of the useful li
of Lake Mead behind Hoover Dam were only about 400 years in anticipation of th
sedimentation process (Smith *et al.* 1960), but as outlined in an earlier chapter it no
appears that some sediment is being stored on valley floors upstream from th
reservoir, and its useful life may be much longer.

The sediments are delivered to the reservoir in one of three ways: flotatior
suspension, or bed load (Schumm 1977; Bogardi 1974). Organic debris, mostly wooc
floats into reservoirs during the spring flood and then is distributed along the shore b
local winds. Suspended sediment is carried great distances into the reservoir from th
mouth of the contributing stream and slowly settles to the bottom to build a wide-sprea
layer of silt and clay. Bed load materials are larger and heavier particles that form
delta extending from the mouth of the contributing stream toward the deepest portion c
the reservoir.

The influence of the reservoir also extends some distance up the contributin
stream as a product of the reduction in the longitudinal gradient. Sediments ar
deposited up the stream as well as down the stream into the delta (Witzig 1944). Thi
reduction of gradient reduces the velocity of flow and sediment transport capability c
the stream. There are several possible models of this upstream distance decay of th
reservoir impact (Figure 14; Bruk and Milorodov 1971; Garde and Swamee 1973
Gessler 1971; Simons and Senturk 1976), but eventually a new equilibrium is estab
lished and this upstream influence is limited to a few kilometers even in the case of th
largest reservoirs (Leopold and Bull 1979; Woolheiser and Lenz 1965).

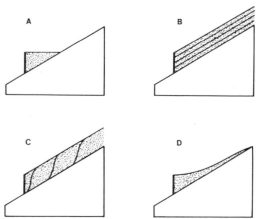

**FIGURE 14**   ALTERNATIVE MODELS OF UPSTREAM SEDIMENT
ACCUMULATION BEHIND DAMS. **A:** accumulation with
a top level with the top of the dam (Miller *et al.* 1962).
**B:** accumulation with sloping bed similar to original channel slope. **C:** backfilling as described by Simons and
Senturk (1976:747-748) and Schumm (1977:152). **D:** a
new profile with a gradient less than the original as proposed by Leopold *et al.* (1964:261).

The alteration of physical processes by dam construction and maintenance results in environmental changes that are significant from the human perspective. The sediments that shorten the life expectancy of the reservoir by filling carry with them chemical compounds that may be hazardous to human health or to fish and wildlife populations. In areas with agricultural activities, pesticides and herbicides are frequently attached to sediment particles, so that the ultimate disposition of the particles is also the ultimate disposition of the chemicals. In natural terrains mineralized zones may contribute heavy metals to the sediment load. Aquatic life in receiving reservoirs accumulate and concentrate the compounds and elements in their tissues to an increasing degree in the higher parts of the food chain. Birds and game fish are especially susceptible as are the humans that consume them (National Park Service 1977a, 1977b, 1977c).

The upstream channel filling associated with reservoir maintenance has two major environmental impacts: flooding and changes in vegetation. The sedimentation causes reduced channel capacities by infilling, increasing the likelihood of overbank flows and endangering near-channel properties. In areas not restricted by narrow canyon walls channel instability is common because of the reduced gradient which promotes lateral migration (general review by Schumm 1977). This situation may be manifested by migrating meandering channels or by braided channels with unstable banks. The newly deposited sediments in the channel and near-channel areas provide expanded seed beds for riparian vegetation, especially phreatophytes (Harris 1966). The massive invasion of species such as willow, tamarisk, and russian olive into areas where they were previously sparse or not found at all profoundly alters local wildlife habitat and coincidentally contributes to the increased flood hazard as outlined previously (Anderson *et al.* 1977; Cohan *et al.* 1978). If reservoir water levels fluctuate, phreatophyte invasion of the shoreline areas is also a potential problem (Warren and Turner 1975).

The newly established shorelines of reservoirs also pose problems of mechanical stability. As rising lake waters inundate terrain that never before experienced aquatic conditions, surface materials become saturated and slopes become unstable (Smith *et al.* 1960; National Park Service 1977c). Rock falls occur on cliffs of sedimentary rocks, and landslides operate to adjust slope profiles to new conditions with saturated materials and footslopes undercut by wave action. In arid regions with sand dunes, dune slides become common at the lake margin.

Lake Mead, the reservoir behind Hoover Dam, provides a useful example of the sedimentation problems associated with large structures (following data from Smith *et al.* 1960). Storage of water began in 1935, and the first time the reservoir was filled with 35 billion cubic m (28.5 million acre feet) of water was in 1941. In the late 1940s an extensive survey of sedimentation in the lake showed that about 68 million kg (150 million tons) of sediment was being poured into the reservoir each year by the Colorado River issuing from the Grand Canyon. Fourteen years after its completion, the sediments reduced reservoir capacity by almost 5%. Although some fine-grained sediments are distributed throughout the reservoir, much of the material was deposited in a huge delta built at the mouth of the Colorado River. In Boulder Basin near the dam, sediment had accumulated to a depth of 43 m (140 ft) in just 14 years, and upstream the maximum thickness of the delta was 82 m (270 feet). When reservoir levels are drawn down for irrigation use downstream and for power generation, extensive mud flats are exposed in the upper portions of the reservoir.

At Lake Mead, upstream impacts on the Colorado River are limited because the waters of the reservoir extend into a steep section of the river's profile, so the sedimentation extends only a short distance above the waters of the reservoir. On the Virgin River, a north-side tributary that flows from southern Utah through the northwest corner of Arizona and southern Nevada, sedimentation extends several miles upstream from the reservoir and has created vast expanses of area susceptible to phreatophyte growth. Suspected water use by these phreatophytes is so great that in 1983 and 1984 the Bureau of Reclamation experimented with removing the phreatophytes and replacing them with crops.

Gillespie Dam, on the Gila River about 16 km (10 mi) north of the town of Gila Bend, Arizona, is much smaller than the high dams on the main Colorado River, but it exemplifies the upstream impacts resulting from sedimentation. The site of the dam is a narrow valley through a Pleistocene Basalt flow, a constriction that was a natural invitation to dam builders. The first structures built by Anglo settlers in the 1870s and 1880s were brush and rock dams rebuilt each year after the spring floods to divert water to irrigated fields a short distance downstream (Granger 1960). By 1892 a masonry dam about 6.5 m (20 ft) high diverted water into two major canal systems, and in 1921 a concrete spanning arch dam about 10 m (33 ft) high was completed. This relatively small structure diverted water for use on more than 33,000 hectares (82,000 acres) of farmland. Fortunately, the purpose of the dam was irrigation diversion and not water storage, because within two years the reservoir area behind the dam was completely filled with sediment.

Once the sediment filled the reservoir area extending about 1 km (0.6 mi) upstream, sedimentation continued to affect the channel in the upstream direction, until by the late 1970s longitudinal profiles clearly showed its impact at least 11.5 km (7 mi) above the dam. This long reach was affected because the constricted area where the dam was built widens into a broad valley immediately above the dam and the stream gradient is quite shallow (unlike the Colorado River above Lake Mead). When tamarisk invaded the Gila River in the early decades of the twentieth century, it flourished on the sediments behind Gillespie Dam and developed into dense thickets (Robinson 1965). These thickets, aided by shallow gradients, accelerated sedimentation in the channel and contributed to its reduced capacity (described for other areas by Hadley 1961; Graf 1978). In the late 1970s and 1980s severe flooding occurred in the areas upstream from the reservoir area, in part because of those reduced channel capacities.

In frustration, area farmers sued the owners of the dam for negligence in its construction and maintenance (*Arlington Improvement Association vs. Northwestern Mutual Life Insurance Company*) but because flooding may have occurred even without the dam, fault was difficult to establish. (The case was settled out of court.) Local and state governments have tried to improve the situation by channelizing the river and cutting a flood path through the tamarisk thicket, but the path cannot be maintained and lateral migration remains a distinct hazard in the area (Figure 11). It appears that near-channel lands will remain at risk.

## Downstream Impacts

High dams influence the downstream river landscape because of radical adjustments made in the hydrologic regime (Leopold *et al.* 1964; Dolan *et al.* 1974). Flood peaks are reduced as a flood control strategy to protect property downstream, but the loss of these flood peaks requires the storage of excess water from flood periods to be released subsequently. These subsequent releases are greater than the discharge of the river would be during non-flood periods, so low flows are increased. High dams that generate electrical power schedule daily releases to insure that higher amounts of water are released through the turbines at times of peak electrical demand. Lower amounts are released during periods of slack demand. Finally, the high dams store almost all of the sediment delivered to their reservoirs and therefore release relatively sediment-free discharges.

In the Colorado River Basin this suite of impacts results in changing a natural river with very large spring floods, low summer flows, and little daily variation of sediment-laden waters to a controlled system with only modest flood peaks in the spring, relatively high summer flows, and drastic daily variation of discharges of clear water. Below Flaming Gorge Dam on the upper Green River, for example, the peak discharge before the dam was about 510 cms (18,000 cfs) but after the dam was built the maximum has only been about 170 cms (6,000 cfs; National Park Service 1977b). Before Glen Canyon Dam affected the Colorado River in the Grand Canyon, median discharge was 210 cms (7,500 cfs) and mean annual floods were 2,440 cms (87,000 cfs), figures which changed to 350 cms (12,500 cfs) and 760 cms (27,000 cfs) after the dam was built (Howard and Dolan 1981). The structure reduced the mean sediment concentration from 1,500 ppm to only 7 ppm.

A short distance downstream from dams, these combined regime changes cause changes in the nature of the river channel by armouring (Harrison 1950; Little and Mayer 1972). In this process, the fine particles on the river bed are entrained by the discharge of the dam because although the flow has significant transport capacity, most of its sediment load has been left behind the reservoir. The transport capacity is filled by materials from the river bed and banks near the dam. The process produces an "armour" on the channel bed because although fine particles are winnowed away, larger cobbles and boulders are usually left in place (Hammad 1972). The erosion process also produces a change in the river profile as erosion occurs near the dam. As downstream distance from the dam increases, the armouring effect and associated erosion decline, but the rate of distance decay is variable, depending on the nature of the dam releases and the channel materials. The armouring effect below Glen Canyon Dam extends only a few km downstream (Pemberton 1976), but below Hoover Dam it reaches over 100 km (60 mi) from the structure (Williams and Wolman 1984).

The armouring effect is a comparatively localized phenomena in comparison to other impacts downstream from high dams (Turner 1971; Fraser 1972; American Society of Civil Engineers 1978). The loss of sediment and changes in regime generally cause a decline in the number and sizes of mid-channel bars or islands and channel-side bars or beaches. Under natural circumstances these features are eroded during spring floods, but sediment-laden lower flows replace the lost material. With a high dam in place, flows are frequently still large enough to cause erosion but the lower flows do not contain enough sediment to replace the losses through deposition. A comparison of photographs taken along the Green River in 1871 with modern conditions shows that these effects are noticeable in Ladore, Whirlpool, and Split Mountain Canyons even though they are 68 km (42 mi) downstream from Flaming Gorge Dam.

Rapids are common features of canyon rivers because the streams have steep gradients to which the rivers adjust by developing a series of alternating pools and riffles (Leopold 1969; Laursen et al. 1976). The locations of the rapids coincide with the sites of rock falls or tributary streams that contribute flash-flood debris to the main channel (Dolan et al. 1978); generally they are spaced randomly along the main stream (Graf 1979b). Before the imposition of high dams, occasional very large floods on the main stream would redistribute the large particles in the rapids. After the reduction in flood peaks by the dams, however, no flows large enough to move the materials in the rapids occur, so the rapids continue to accumulate particles. In a 15-year sample period from 1961 to 1976 five major rapids experienced change in the Colorado River Basin on the Colorado, Green, and Yampa rivers (Graf 1980b). In every case the rapids increased in size or were newly created.

Physical changes in the fluvial system are reflected in the riparian ecological community (Ward and Stanford 1979). Vegetation systems, that under natural conditions were adjusted to drastic water level fluctuations, must change to accommodate more consistent and generally higher flows (Figure 15). In many cases these changes result in greater species diversity as more clearly defined ecological niches are formed based on distance to the water's edge. Usually the density of growth is greater in the post-dam condition. Exceptionally dense growth of phreatophyte species replaced nearly barren ribbons of sand downstream from Coolidge Dam (Gila River, Arizona) and Navajo Dam (San Juan River, New Mexico). These new vegetation communities give rise to changes in wildlife as new populations take advantage of changes in habitat. In riparian communities of the Colorado River Basin, avian populations are especially susceptible to change. Fish also respond to upstream dams because of the change from high silt loads to more clear, colder water (Bolke and Waddell 1975). The development of trout fisheries below dams is a common feature of the main stem of the Colorado River.

The physical and ecological changes downstream from dams, combined with rapidly increasing demands for whitewater recreation, have forced environmental managers to regulate the human use of all the major canyons in the Colorado River system (Table 5; Figure 16; National Park Service 1977a, 1977b). The National Park Service, Forest Service, and Bureau of Land Management impose quotas on the number of people who can use the rivers and they enforce strict regulations with regard to areas that can be used for camping, where and how fires are to be used, and the management of waste materials. With the reduction in the capability of the rivers to erode polluted beaches and then replace them with clean material, control of human impact takes on increasing importance. Because of the increased severity of rapids, only licensed and experienced pilots are permitted to conduct river tours, and private individuals must pass competency tests before they are permitted to "run the canyons."

The Grand Canyon probably best typifies the downstream impact of dams (following data from Turner and Karpiscak 1980; National Park Service 1977c). Glen Canyon Dam was completed in 1962 immediately upstream from Grand Canyon and its impacts have become apparent in the subsequent 20 years. Many channel-side bars and beaches have been seriously eroded and reduced in size, while others have enlarged. During the largest releases of the dam (in 1983) some beaches were enlarged and most were at least "cleansed" with some additions of sand, but over the two decades since the dam closure, there has been a net loss of beach space. In the twenty years since the dam was closed, an equilibrium condition has not been established. In 1974 27-Mile Rapid was created by a flash flood on tributary Tiger Wash (Graf 1979b), and in the early 1980s Crystal Rapid became increasingly severe with the addition of new tributary material. The discoverer of the newly enhanced rapid drowned when his raft was upset in the unsuspected turbulence.

Riparian vegetation communities have completely changed during the post dam period (Figure 15). Before the dam, riparian vegetation was largely absent or consisted

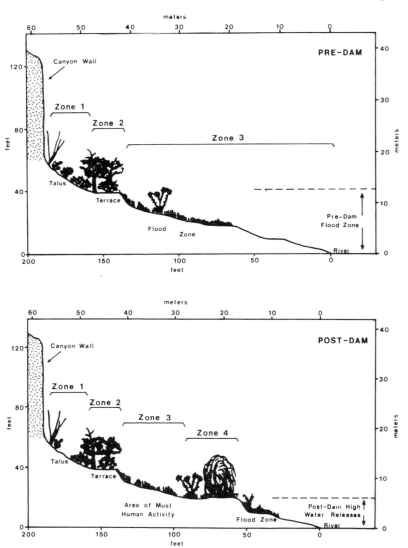

**FIGURE 15** PRE-DAM AND POST-DAM RIPARIAN VEGETATION IN THE GRAND CANYON DOWNSTREAM FROM GLEN CANYON DAM. Vegetation zones: 1, stable desert vegetation; 2, stable woody vegetation; 3, unstable zone; 4, new riparian vegetation, phreatophytes. From National Park Service (1979), redrawn from original by S. W. Carothers.

of only sparse cover. After the dam, reduced variation in water levels have resulted in more dense communities including tamarisk (*Tamarix chinensis* Lour.), camelthorn (*Alhagi camelorum* Fisch.), sandbar willow (*Salix exigua* Nutt.), desert broom (*Baccharis sawthrides* Gray), and cattail (*Typha* spp.). Elimination of the highest flood peaks permitted the establishment of Russian olive (*Elaeagnus agustifolia* L.) and elm

**TABLE 5**    THE WHITEWATER RECREATION RESOURCES OF THE
COLORADO RIVER BASIN

| Map Key | River | Canyon | State | Agency |
|---|---|---|---|---|
| 1 | Colorado | Below Laguna Dam | AZ/CA | BIA |
| 2 | Colorado | Blyth to Imperial | AZ/CA | BLM |
| 3 | Colorado | Grand | AZ | NPS |
| 4 | Colorado | Cataract | UT | NPS |
| 5 | Colorado | Fisher Towers Area | UT | BLM |
| 6 | Colorado | Westwater | UT | BLM |
| 7 | Colorado | Loma to Westwater | CO | BLM |
| 8 | Colorado | Glenwood Canyon | CO | FS |
| 9 | Colorado | Pumphouse to Glenwood | CO | BLM |
| 10 | Green | Stillwater | UT | NPS |
| 11 | Green | Labyrinth | UT | NPS |
| 12 | Green | Gray | UT | BLM |
| 13 | Green | Desolation | UT | BLM |
| 14 | Green | Split Mountain | UT | NPS |
| 15 | Green | Whirlpool | UT/CO | NPS |
| 16 | Green | Ladore | CO | NPS |
| 17 | Green | Red | UT/WY | BLM/FS |
| 18 | Gila | The Box | AZ | BLM |
| 19 | Gila | Upper Gila | AZ/NM | BLM/FS |
| 20 | San Francisco | Upper San Francisco | AZ/NM | BLM/FS |
| 21 | Salt | Upper Salt | AZ | FS |
| 22 | Verde | Upper Verde | AZ | FS |
| 23 | Bill Williams | Lower Bill Williams | AZ | BLM |
| 24 | Animas | Animas | CO | BLM/FS |
| 25 | Dolores | Dolores | CO/UT | BLM/FS |
| 26 | Gunnison | Black | CO | NPS |
| 27 | Yampa | Yampa | CO | NPS |

Data from Whitewater Committee (1982). For locations refer to Figure 16.

(*Dyssodia* supp.). Also, the Colorado Squawfish, a gargantuan relative of the minnow that occasionally grew to lengths of 2 m (6 ft) and now an endangered species, has largely disappeared from the river (National Park Service 1977c). The several km reach of the Colorado from the dam to Lee's Ferry is now dominated by clear cold water released from the dam and has become the most heavily fished trout habitat in Arizona.

Visitors to the Colorado River in the Grand Canyon became so numerous that the administrator of the canyon, the National Park Service, has developed an elaborate management plan to accommodate the annual average of 15,000 river runners (before 1950 fewer than 100 people had gone through the canyon in its entire history: Goldwater 1970). A precise schedule is established for boaters who must depart the starting point (Lee's Ferry) and utilize designated camping sites on a rigid timetable so as to avoid overcrowding and preserve at least some semblance of a wilderness river experience. Fires must be built only on metal pans to protect the quality of the sand beaches, and all waste materials must be carried out of the canyon. The significance of the river-running industry in the Grand Canyon is reflected in changes of the management plan forced on the park service directly by a lobbyist-sensitive congress (Larson 1974), and in the $20 million of economic activity it generates each year.

The high dams of the Colorado River Basin provide security and wealth for the southwestern United States and are directly responsible for much of the development presently enjoyed by inhabitants and visitors to the region. However, these benefits of controlling some aspects of the river system have come at the cost of increased instability of other parts of the system. In some cases expectations of benefits were too high; no economically feasible system of high dams can impose complete control on a natural river system. Society must come to terms with the fact that although the river is a priceless resource, the Colorado is still a hazard to be regarded with care, a delicate natural system susceptible to undesirable side effects from the application of technology.

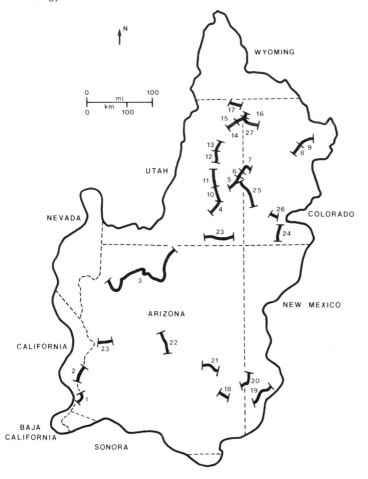

**FIGURE 16**   WHITEWATER RECREATION RESOURCES IN THE COLORADO RIVER BASIN. Numbers keyed to Table 5.

# 7

# Chemical Instability

Environmental instability and change in the Colorado River Basin has chemical as well as mechanical aspects. Effective human use of the surface environment depends on knowledge and predictability of the chemical characteristics of the soil and water. However, during the last half century understanding of chemical cycles in natural circumstances has only begun to develop, and several remarkable chemical changes have occurred that pose significant environmental management problems. The following chapter outlines some of the major chemical changes with regard to salinity, heavy metals, and radioactive materials, and explores the human implications of the changes.

## Salinity

Salinization, the increasing saltiness of water and soil, is a major management problem associated with irrigation agriculture (Casey 1972). Water that is transpired by growing plants and evaporated from soil or water surfaces leaves behind its dissolved minerals, increasing the concentration of materials in the remaining water (Figure 17: Gardner and Stewart 1978). Under natural conditions the salt concentration in Colorado River water in the lower reaches probably ranged from 250 ppm (Iorns et al. 1965) to 380 ppm (Sheridan 1981). Natural sources of salt in the form of saline springs empty small amounts of water into the river but account for almost half of the salt load (Evans 1975). In several places, major streams of the basin flow over exposed salt beds, two of the most obvious being near Canyonlands National Park where the Green River flows over Epsom salts and in eastern Arizona where the Salt River flows over sodium chloride beds so prominent that they lend their name to the river (Granger 1960).

By the 1970s, with thousands of acres in irrigation, 12% of the river's flow evaporating from reservoirs, and 3% being diverted to other basins, the salt concentration ranged from 50 ppm in headwaters to over 800 ppm at Imperial Dam near Yuma, Arizona (Holburt and Valentine 1972). For comparison, the U.S. Department of Agriculture defines 750 ppm as a high risk level for agriculture, and the U.S. Environmental Protection Agency specifies 500 ppm as the upper limit for human consumption (Gardner and Stewart 1978). Under natural conditions, the salt concentration in the Colorado River at Lee's Ferry was about 250 ppm, but by 1957 it had doubled. Much of the increase resulted from irrigation in the Upper Basin where each irrigated acre contributed an additional 1.7 tons (0.38 kg/sq m) to the salt load of the river (Holburt and

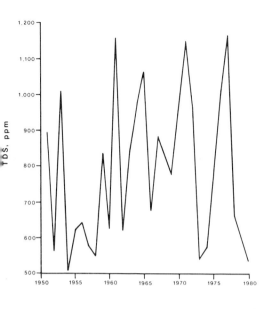

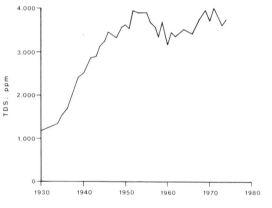

**FIGURE 17** DISSOLVED SOLIDS IN THE GILA RIVER, ARI-
ZONA. Above: dissolved solids in surface water of
the Gila River near Kelvin, Arizona, showing substantial fluctuations in response to variability in total
water discharge which dilutes the solute load.
Below: dissolved solids in shallow groundwater
near the Gila River, Buckeye, Arizona, showing the
generally increasing solute load resulting from long-
term irrigation, infusion of salt-rich drain water, and
depletion of relatively salt-free water upstream. Unpublished data from Water Resources Division, U.
S. Geological Survey.

Valentine 1972). Evaporation from Upper Basin reservoirs increased the salinity of the river by about 17 ppm at Lee's Ferry, while similar losses in the Lower Basin hiked the salinity 80 ppm at Imperial Dam.

The consequences of this increased salinity of the Colorado River are increased costs in the use of the water resource and declines in agricultural productivity (Casey 1972). Saline waters destined for urban and industrial use must be treated (desalinized) to reduce the salt content, and deposition of the salt in fields renders them useless for many crops. Salt-tolerant crops may be substituted in some cases, but sometimes the fields must be abandoned; they may not be simply flushed with clean water because drainage of the waste water is frequently a problem. Finally, because Mexico lies at the downstream end of the river, the 1.5 million acre feet promised by treaty to that nation has become water of increasingly poor quality (Brownell and Eaton 1975). The result has been a debate of mammoth proportions concerning the geographic division of responsibility for salinity problems and their resolution.

Before 1961 Mexico received water with about 800 ppm salt, barely usable for irrigation. In 1961-1962 two alterations of the river system occurred (Cabrera 1975; Furnish and Ladman 1975). First, pumps began moving salt-rich water from the Wellton-Mohawk Irrigation District along the lower Gila River into the Colorado River. Second, the flow of the river was drastically reduced as Glen Canyon Dam was closed and Lake Powell began to fill. The result was that Mexico received only the 1.5 million acre feet of water agreed to by treaty (before, the average annual total was over 4 million acre feet) and the water contained an average of 1,500 ppm of salt with maximum values as high as 2,700 ppm. Water quality was so poor that Mexican farmers simply conducted it to the Gulf of Baja California rather than risk ruin to their soils by using it.

The major culprit in this increased salinity was the waste water from the Wellton-Mohawk Irrigation District with its salinity in the range of 6,000 ppm (Sheridan 1981). A canal was constructed to conduct it to the sea, but this water was then lost to the accounting system of the basin and had to be replaced from shares of the basin states to meet commitments to Mexico. In 1973 Mexico and the United States agreed that the U.S. would deliver water to its southern neighbor with a salinity no more than 115 ppm greater than that at Imperial Dam, a short distance upstream from the border. Without major multi-million dollar efforts to solve the Wellton-Mohawk salinity problem, the U.S. will not even approach this goal.

The two primary routes to solution of the salinity issue are structural and non-structural. The construction of a desalinization plant near the international border will cost at least $333 million and $12 million per year to operate (U.S. Bureau of Reclamation 1975). The costs will be borne by the federal government, so that the nation as a whole will support benefits for a few in the Wellton-Mohawk area, the justification being that the expenditure is the result of an international agreement entered into by the federal government. Non-structural alternatives include halting the flow of irrigation water into the Wellton-Mohawk district and retiring the land from production. Though economically more efficient in the long run (total cost about $550 million to buy up the farm lands), this localized approach to the problem source is politically unacceptable: local land owners do not want to cooperate and the federal government will not force the issue. The result is a huge national investment in a structural effort to affect the chemical characteristics of the river.

## Heavy Metals

Gold and silver attracted many prospectors to the Colorado River Basin during the final decades of the nineteenth century. These and subsequent mining and milling operations involving heavy metals profoundly impacted on the chemistry of the river system (for general principles, see Fortescue 1980; Levinson 1980). Copper has been the most prominent metal in economic terms, but vanadium, radium, and uranium have also been mined and milled in large quantities. Mining and milling processes release large amounts of these heavy metals which are then mobilized in the surface environment. Occasionally they become concentrated at levels that threaten wildlife and plants or that endanger human health.

Heavy metals travel in the river systems as solutions and as particles entrained by the flow (Wolfenden and Lewin 1978). Most heavy metals occur in the rocks of the basin in chemical compounds that dissolve readily in water, so they move in ground and surface waters (Wedepohl 1978). In milling operations, the metals are concentrated in liquid wastes that must be carefully contained to prevent pollution of groundwater by seepage. Water discharged from mines introduces high concentrations of dissolved metals to surface flow. Most heavy metals are also transported by streams as particles on the bed of the channel either as compounds or in elemental form (Wolfenden and Lewin 1978). Life forms in the channel or in reservoirs further concentrate the heavy metals in their tissues, sometimes to hazardous levels (Summerfield *et al.* 1976). This process of "bioamplification" results in multiplying the concentration of some metals by more than 20,000 times by the time the material reaches the top of the food chain (Sittig 1981).

In spatial terms, the impact of mining and milling on chemical characteristics of rivers declines exponentially with distance (Lewin *et al.* 1977). Most studies of heavy metal distributions have been conducted in humid regions, but Marcus (1983) investigated a typical arid/semi-arid case in central Arizona. Tailings in the town of Superior have released copper particles and compounds into Queen Creek for several decades, with the resulting elevated levels of copper that declined logarithmically to a point several km downstream. Copper concentrations ranged from 15 to 4,400 ppm (the desirable limit in drinking water for safe human consumption is 1 ppm: U.S. Environmental Protection Agency 1976). The copper concentrations were highest in the fine sediments (a commonly found arrangement: Huff 1970), and were inversely related to drainage basin area and diluting sediment yields. The major exception to this well-defined spatial structure was dilution of the copper concentrations by masses of sediment deposited in the main channel of Queen Creek by flash floods in tributaries.

Human impact on chemical cycles of the basin is not limited to small scales. The movement of mercury in the Upper Basin, for example, has been completely disrupted by Glen Canyon Dam (Standiford *et al.* 1973). Mercury occurs in the form of dispersed mercury chloride in many of the rocks exposed in the Colorado Plateau portion of the Upper Basin and is released during weathering (Cadigan 1970). After a series of chemical changes the mercury appears in the sediments of the region's streams in the form of elemental mercury or in the form of relatively insoluble salts (Levinson 1980). It is frequently attached to the fine sediments that are readily carried by the streams, and eventually is deposited on the floor of Lake Powell. Biota on and near the bottom sediments ingest the mercury, which is increasingly concentrated in higher life forms. Bass from the lake have concentrations of over 700 ppb of mercury in their tissues

(National Park Service 1977c), a hazard to human consumers of the fish since the Environmental Protection Agency's recommended limit for human consumption to avoid damage to the central nervous system is only 500 ppb (U.S. Environmental Protection Agency 1976; Sittig 1981). The lake is a sink for mercury because the sediments carrying the metal in an insoluble form are trapped there, and water releases do not carry the sediments or the mercury further downstream (Potter *et al.* 1975).

The regional mercury budget for the upper Colorado River Basin shows the remarkable variability in mass budgets for a typical semi-arid river system (Figure 18). The Colorado River is the primary source of water for the lake, but it is relatively devoid of sediment and mercury because its source areas in the Rocky Mountains are dominated by relatively high precipitation and crystalline rocks. At the other extreme are the local tributaries (Canyon Lands streams drain the arid Colorado Plateau with easily eroded, mercury-rich materials). Very little water (only 9% of the total) comes from these surfaces, yet they contribute disproportionate shares of the sediment and mercury (40% and 36% of the respective totals). Mercury is therefore a natural component of the Upper Basin environment, and its inflow to the lake is not likely to be controlled. A management solution begun in 1984 is the introduction of fish species which do not concentrate mercury in the parts of their bodies consumed by humans.

### Radioactive Materials

The upper Colorado River Basin has been a significant source of radioactive ores since the early 1900s when Madame Curie travelled to western Colorado and eastern Utah to select ores for her radium experiments in France (Gray 1982). Until the end of

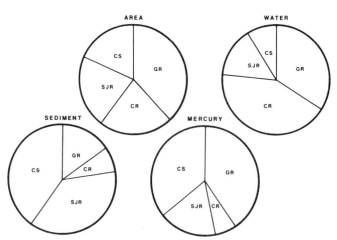

**FIGURE 18**    MASS BUDGETS FOR LAKE POWELL AND THE UPPER COLORADO RIVER BASIN. Note that the contributions to various parts of the budget are not directly related to area, indicating that the geography of water is different from the geographies of sediment and mercury. GR, Green River; CR, Colorado River; SJR, San Juan River; CS, Canyon Lands streams.

World War II, uranium ores, mostly carnotite, that were found in the region during mining operations for other metals were discarded as waste. After 1945, however, the national mandate to develop uranium sources for atomic weapons lead the Atomic Energy Commission to offer price supports and bonus grants for new uranium discoveries. A prospecting and development boom not unlike the California gold rush almost a century earlier occurred in the Upper Basin, with the development of hundreds of small uranium mines, several large ones, and more than a score of processing mills (Taylor and Taylor 1970). The boom collapsed in 1957 when the Commission determined that its uranium supplies were adequate and that it would no longer provide financial incentives. In the subsequent three decades the uranium market and its associated mining and milling activities have waxed and waned in the basin several times in response to foreign marketing activities, economic cartels, and the vacillations of the nuclear power industry (Perkins 1979).

As a result of local mining and milling activities, radioactive materials have been mobilized in the surface environment of the Colorado River Basin through the atmosphere, water, and surface materials. Radioactive radon gas escapes from open, abandoned mine shafts in quantities that may threaten human health in the surrounding area (Wilkening *et al.* 1972). Many of the most productive mines are sunk in aquifers and must be continuously de-watered by pumping water from the mine to the surface, producing potentially hazardous conditions (U.S. Environmental Protection Agency 1975). This water with its associated radioactivity is emptied into surface water courses that are thus converted from small ephemeral streams to larger perennial ones.

Perhaps the most difficult problem with radioactive materials concerns tailings, the solid waste materials produced by milling activities (Nelson and Shepherd 1982). During the milling process, uranium ore is crushed and mixed with water along with acids, salts, and a variety of noxious chemicals (Merritt and Pings 1969, 1971). The uranium chemically combines with resins to form "yellowcake," a concentrate that is shipped out of the region for further processing. The tailings that make up the waste materials in this process (about 99% of the ore that is mined) are a mixture of liquids and solids that are stored in ponds on the surface. The tailings contain uranium not recovered during the milling plus thorium, radium, and other elements depending on the source of the ore. The management of these tailings represent a difficult problem in preserving the environmental quality of the region because they may seep into local groundwater supplies or when the materials dry they may be widely distributed by wind (Kaufman *et al.* 1976; Clements *et al.* 1978). One tailings pile was located near Hite on the Colorado River and is now drowned by Lake Powell, with the fate of its materials unknown. Surface erosion of the tailings by running water may carry them into streams and distribute them throughout the drainage system.

The spatial aspects of the tailings management problem are illustrated by the example of the Puerco River, a northwest New Mexico tributary to the Little Colorado and Colorado rivers. (Present practice is to use the partly Anglo name for the stream — which is also known as the Rio Puerco del Oeste — to avoid confusion with the Rio Puerco, a tributary of the Rio Grande near Albuquerque). In 1979 a tailings pond dam burst near Church Rock, New Mexico, releasing into the nearby Puerco River almost 100 million gallons of liquids and 1,100 tons of sediment (Kunkler 1979). The flood wave of tailings liquid and subsequent natural floods distributed the radioactive materials more than 40 km (25 miles) downstream where they may have affected the

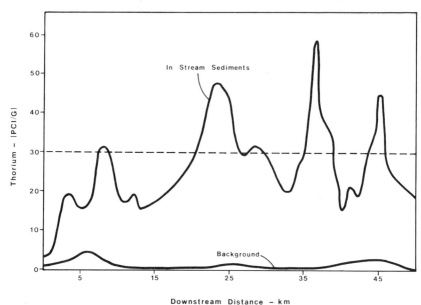

**FIGURE 19**    DOWNSTREAM DISTRIBUTION OF THORIUM-230 IN THE PUERCO RIVER BELOW THE TAILINGS SPILL AT CHURCH ROCK, NEW MEXICO. The safe level limit as recommended by the Environmental Protection Agency is shown by the dashed line. Data from Weimer *et al.* 1981.

quality of surface environments. As the liquids percolated into the porous bed of the stream it is also possible that shallow aquifers were contaminated.

Although the initial expectation might have been to observe an exponential decay of concentration of radioactive materials with increasing distance from the site of the spill, this manifestation of the geographic version of the gravity law did not appear (Figure 19; Weimer *et al.* 1981; Wolfenden and Lewin 1978). Apparently the concentration of radioactive substances varies in the downstream direction because the materials are moving as masses or slugs of sediment (Pickup *et al.* 1983), rather than in a smoothly entrained and distributed whole. The movement of sediments in pulses in arid-region rivers may be a natural reflection of the flashy nature of their discharges or of discontinuous contributions from tributaries (Schumm 1977). As a result, the level of radioactivity in the Puerco River alternates in the downstream direction between relatively low values and values that are high enough to exceed the safe limits of 30 picocuries per gram.

Environmental instability and change in the Colorado River Basin therefore are not limited to physical dimensions, but are also an issue in the less obvious, but no less important, chemical realm. Each development project in the basin has a hidden cost in terms of its impact on the regional chemical cycles that is often not even suspected until after the project is completed. The real question facing society is not whether the development projects will go forward; popular political pressure for them is so great that opposition to major projects will not prevent them. The real question is which of several alternative choices will be pursued.

# 8

# Futures for the Colorado River Basin

A prediction of the precise course of natural and human affairs in the Colorado River Basin is not possible given our present understanding of processes, but certain trends appear to be well-established and are likely to continue at least into the twenty-first century. The following chapter begins with the indentification of the trends most likely to have environmental significance for the basin and then highlights some identifiable management approaches that are developing to accommodate those trends. Finally, the role of geographers in the management of the basin environment is outlined with examples.

## Significant Trends

For several decades the concept of carrying capacity has been a common management tool for range and wildlife managers, but it has only been in the last twenty years that the concept has been applied explicitly to people (Hendee *et al.* 1978; Flader 1974). How many people can the Colorado River Basin support given our present and foreseeable technology is the major unanswered environmental question for the region, a question that becomes increasingly important given the growth of population in the basin states. Migration from the American Midwest to the Southwest and Sunbelt (Biggar 1979; Fitzsimmons *et al.* 1980), along with high birth rates among minority populations, have made the Colorado River Basin states of Nevada, Arizona, New Mexico, and Colorado among the fastest growing states in the nation, an unrelenting upward pressure on the rates of use of already scarce resources.

The population in the basin also is likely to continue a well-established trend toward increasing urbanization (Hiibner 1974). The basin states have consistently been among the most urbanized in the nation throughout their history, but they are continuing to concentrate even more of their populations in urban places (Pickard 1973). The attending demands by urban places on surrounding open space, recreational resources, water supplies, and air sheds for the dissipation of air pollution, are also likely to increase.

Population growth and urbanization are accompanied in the Colorado River Basin by economic development which has important environmental implications. The early development of many cities in the basin relied on the mining of groundwater, that is, the withdrawal of groundwater reserves at rates greater than natural replacement rates. Declining water tables will force a change from reliance on groundwater to reliance on surface water. Such a change may in part be accommodated by conservation measures (Martin *et al.* 1984), but it also requires massive investments and large construction projects in fragile arid-region ecosystems: the Central California, Central

Arizona, Central Nevada, and Central Utah projects are examples. The cost of the newly developed water supply is high in environmental as well as financial terms.

The conversion of irrigated agricultural land to urban or suburban land is not likely to result in a strain on existing developed water supplies in the southern portion of the basin. In the Salt River Valley, most irrigators use an average of four to five acre feet of water per acre per year, an amount similar to the quantity required by occupants of residential neighborhoods that are replacing the farmers. Expansion of the urban areas into previously unirrigated zones, however, may prove to be impossible on a large scale. The future geography of the region's cities may therefore in part be forecast by the present geography of the region's agriculture.

The social and economic development of the basin will surely take place against a backdrop of continuing environmental change. The variability in rainfall and runoff outlined in earlier chapters will probably continue, but whether or not we will soon return to the exceptionally moist conditions of the early twentieth century remains to be seen. Conditions in the early 1980s provide us with a glimpse of what the situation was like then, but the persistence of the atmospheric circulation patterns is unknown. The instability of the vital channel and near-channel environments is likely to continue, with alternating periods of erosion and sedimentation that must be viewed as normal rather than as unusual adjustments in an otherwise static and predictable condition.

The chemical stability of the basin's atmosphere, water, and soils will continue to be radically altered on a local basis with important but often unforeseen implications for human health. The most recent problems, for example, have centered on the dumping of carcinogenic industrial compounds into deep injection wells where they adversely affect groundwater supplies. The effects are being discovered in some cases thirty years after the original disposal.

## Management Perspectives

The environmental management strategies that American society has designed to cope with these problems are undergoing major changes in the 1980s. In the case of water development projects, there is a clear impending shift from federally sponsored projects to state-sponsored efforts. Although this shift is in part a reflection of the political climate of a nation attempting to limit federal involvement and investment in regional affairs, it is also in part a reflection of the nature of the resource. After nearly eighty years and $1.7 billion of construction, the federal government has developed all the most effective dam sites in the basin. The remaining sites are largely confined to rivers in national parks or are designated as wild and scenic rivers and therefore are not available. These sites enjoy the implicit protection of a strong national political lobby that has as its objective preservation as opposed to development (according to its 1983 financial report, the Sierra Club alone boasts an $18.4 million budget and a membership of more than 300,000). Therefore, smaller state funded projects are likely to be more common in the next few decades than they have been in the past.

By bringing the costs of development and flood protection closer to the users in the states affected, the already visible shift from structural to non-structural approaches is likely to accelerate. Investments in massive flood-control structures are likely to be replaced by stronger and more frequently enforced zoning restrictions designed to prevent development in hazardous areas. This policy change is bound to come slowly to the basin states, however, where private property rights are strongly defended.

Environmental management strategies are likely to be strongly influenced by inter-regional conflicts. The continued export of electrical power from the basin and the importation of pollution (in the case of coal-fired power plants) is re-enforced by stiffening regulations elsewhere that are not matched by regulations in most basin states. The regional exchange even extends to nuclear energy. In the middle 1980s the nation's largest nuclear generating station will be the Palo Verde station in central Arizona, but the majority of its electricity will be exported to California.

Inter-regional conflicts with regard to water have been a feature of the basin's developmental history from the beginning, and they are likely to continue (Fradkin 1981; Waters 1946). As the southern basin continues to grow in population and economic activity, the amount of available Colorado River water will be inadequate for an assured supply. Augmentation of the river's flow by artificially stimulated precipitation or massive transfers of water from other basins offer the only possible solutions, but in each case regional political barriers appear insurmountable. If precipitation is stimulated in the Colorado River Basin, will this rob downwind states of the moisture they otherwise would have received under natural conditions? And if water is to be transferred from another basin into the Colorado Basin, from whose area will it come? This question is so serious that authorization of the Central Arizona Project was dependent on an agreement between congressional representatives from the Southwest and those from the Pacific Northwest that interbasin transfers would not even be discussed for at least a decade.

## The Role of Geographers

Geographers are taking an active role in dealing with the management of the Colorado River Basin by their activities in the major agencies that deal with the issues outlined above. At a local level human geographers are commonly found in positions in county or city planning agencies that attempt to direct the development resulting from population growth. Geographers are found in planning agencies in every basin state. Resource geographers are employed by state land and water agencies to assess the available resources and direct their management. (The Director of the Utah State Lands Department for a number of years was a geographer.) Cartographers and remote sensing specialists are employed at all levels of government to aid in the inventory and mapping of natural and cultural resources (at the National Park Service Western Regional headquarters in Denver, for example).

Physical geographers commonly deal with the management problems associated with water or land. The Corps of Engineers is commonly thought of as employing only engineers, yet local offices in some cases have numerous geographers (Phoenix and Los Angeles, for example), and the Los Angeles District Office which administers the Lower Basin has a position entitled "Chief Geographer." The Forest Service, Park Service, and Bureau of Land Management employ geographers in the preparation of environmental impact assessments for projects and management plans.

Human and physical geographers are employed by private companies and semi-public agencies in dealing with environmental management. Public utilities for power and water require impact assessments, demographic predictions, and economic analyses that result in their employment of geographers. Private consulting firms have the same requirements, as well as cartographic and remote sensing needs. Legal firms dealing with environmental and resource matters employ geographers as consultants and advisors.

Geographers are hired for these positions because they bring to the problem of environmental management a particular perspective. The geographer adopts a system-wide view of issues and deals with problems not as isolated occurrences but as part of a complex process-response or control system. The public perception of geography is poorly defined, probably in part because geographers address so many different questions, but that diversity is also the geographer's strength in dealing with environmentally-related problems. Geographers frequently fill coordinating positions, especially in the development of environmental impact assessments, because geographers can communicate effectively with contributors ranging from engineers to sociologists.

Balancing this rather eclectic world view, however, is the major distinguishing characteristic of geographers: a spatial perspective. The problems of dealing with the instability of the environmental systems of the Colorado River Basin are inherently spatial problems: the distribution of water, land, and people. As shown in previous chapters a key to understanding the instability of the environmental system is to understand how the elements of those systems are distributed on the surface of the earth and why those seemingly dependable distributions sometimes change. Effective solutions to the problems therefore must also be inherently spatial, and geographic scientists, by their perspective and training, have much to contribute.

An example of the contributions of a geographer in the management of the Colorado River Basin is provided by an announcement of a vacancy by the Corps of Engineers for a B.S. or B.A. degree holder to fill a position labelled "Geographer" in the Phoenix area at an annual salary of $25,300 (Office of Personnel Management, Announcement FP-4-29, April 16, 1984). The description of duties was as follows:

> The incumbent of this position develops, coordinates, and carries through to completion geographic studies and research, including long term (over one year duration) complex studies that cover large geographic areas (entire drainage basins, *e.g.*, 2,000 sq mi). In this connection identifies, collects, evaluates, coordinates, synthesizes, analyzes, interprets and presents a wide variety of geographic data including climatology, geology, geomorphology, soils, hydrology, hydraulics, land use, mobility, transportation, economics, demography, and historical flood phenomena, as related to flood control and other aspects of water resources planning in formulating recommended projects and alternatives. Integrates geographic information from a regional and topical viewpoint for presenting and illustrating in narrative and graphic form. Determines causal relationships between various factors as related to proposed project plans.

## Conclusions

The implications of environmental instability for the management of the Colorado River Basin are that a harmonious relationship between society and the basin's environmental resources is most likely to develop if society adopts an accommodating approach. The application of technology can increase the quality of life for society's members and can make possible economic development in adverse environments. The application of technology cannot remove all uncertainties (Rhodes *et al.* 1984), however, and inevitably results in additional costs.

Technology, whether structural or non-structural, must also be tailored to the environment in which it is applied. Many of the costs of the development of the Colorado River Basin have hidden costs in terms of environmental damage or the creation of hazardous situations that result from the application of technology developed in humid regions. American society's conceptions of how environmental systems, especially river systems, behave is largely predicated on experience in humid regions. Engineering functions developed for canals do not apply effectively to arid-region rivers, laws developed for clearly defined channels in humid regions do not provide useful regulations in arid areas, and planning processes based on the relatively stable, long term averages found in eastern environmental systems do not address the highly changeable western conditions.

The solution to these dilemmas and the general lesson of environmental management in the Colorado River Basin are embodied in a culture that existed there a thousand years ago. Whether Kokopelle really existed or whether he was a widely held myth is not important. Present day wanderers in the region may imagine that they can still hear notes from his flute echoing off the rock walls, but it is only the song of a canyon wren. The man (or his spirit) represented the concept that although society can alter environments to meet its needs, it will survive longest and be most prosperous if it accommodates itself to the environment instead of striving for complete control. Adaptation rather than domination is not an especially common trait for Western cultures, but long-term habitation of the Colorado River Basin may depend on such a philosophy.

*Kokopelle*

# Bibliography

**Aldridge, B. N. 1970.** *Floods of November 1965 to January 1966 in the Gila River Basin, Arizona and New Mexico, and Adjacent Basins in Arizona.* U. S. Geological Survey Water-Supply Paper 1850-C.

**Aldridge, B. N. and Eychaner, J. H. 1984.** *Floods of October 1977 in Southern Arizona and March 1978 in Central Arizona.* U. S. Geological Survey Professional Paper 2223.

**Allin, C. W. 1982.** *The Politics of Wilderness Preservation.* Westport, CT: Greenwood Press.

**Ambler, J. R. 1977.** *The Anasazi.* Flagstaff, AZ: Museum of Northern Arizona.

**American Quaternary Association. 1976.** *Guidebook, San Pedro Valley Field Trip, October 11, 1976.* American Quaternary Association, Fourth Biennial Meeting, Arizona State University, Tempe.

**American Society of Civil Engineers. 1978.** *Environmental Effects of Large Dams.* New York: American Society of Civil Engineers.

**Anderson, B. W. *et al.* 1977.** "Avian Use of Saltcedar Communities in the Lower Colorado River Valley," pp. 128-136 in R. R. Johnson and D. A. Jones (editors), *Importance, Preservation and Management of Riparian Habitat: A Symposium.* U. S. Department of Agriculture, Forest Service General Technical Report RM-43.

**Antevs, E. 1952.** "Arroyo Cutting and Filling," *Journal of Geology* 60:375-385.

**Arnold, L. W. 1973.** *A Study of the Factors Influencing the Management of and a Suggested Management Plan for the Western White-Winged Dove in Arizona.* Phoenix, AZ: Arizona Game and Fish Department.

**Atherton, L. 1961.** *The Cattle Kings.* Lincoln, NE: University of Nebraska Press.

**Barry, R. G. *et al.* 1981.** "Synoptic Climatology of the Western United States in Relation to Climatic Fluctuations During the Twentieth Century," *Journal of Climatology* 1:97-115.

**Bartlett, R. A. 1962.** *Great Surveys of the American West.* Norman, OK: University of Oklahoma Press.

**Baum, B. R. 1967.** "Introduced and Naturalized Tamarisks in the United States and Canada," *Baileya* 14:19-25.

**Baumann, D. D. and Kates, R. W. 1972.** "Risk From Nature in the City," pp. 169-194 in T. R. Detwyler and M. G. Marcus (editors), *Urbanization and Environment,* Belmont, CA: Duxbury Press.

**Baumann, N. and Elmer, R. 1976.** *Flood Insurance and Community Planning.* University of Colorado, Institute for Behavioral Studies, Natural Hazards Working Paper #29.

**Berkman, R. L. and Viscusi, W. K. 1973.** *Damming the West.* New York: Grossman Publishers.

**Biggar, J. C. 1979.** *The Sunning of America: Migration to the Sunbelt.* Population Reference Bureau, Population Bulletin 34, 1.

**Bogardi, J. L. 1974.** *Sediment Transport in Alluvial Streams.* Budapest, Hungary: Akademai Kaido.

**Bolke, E. L. and Waddell, K. M. 1975.** *Chemistry and Temperature of Water in Flaming Gorge Reservoir, Wyoming and Utah, and the Effect of the Reservoir on the Green River,* U. S. Geological Survey Water-Supply Paper 2039-A.

**Borchert, J. R. 1950.** "The Climate of the Central North American Grassland," *Annals, Association of American Geographers* 40:1-39.

**Borland, M. 1971.** Chapter 29, in H. W. Shen (editor), *River Mechanics.* Fort Collins, CO: Colorado State University.

**Bowser, C. W. 1960.** "The Phreatophyte Problem," *Proceedings of the Arizona Watershed Symposium* 4:17-18.

**Brown, C. D. 1944.** "Discussion of 'Sediment in Reservoirs' by B. J. Witzig," *Transactions of the American Society of Civil Engineers* 109:1080-1086.

**Brown, D. E. et al. 1977.** "Inventory of Riparian Habitats," pp. 10-13 in R. R. Johnson and D. A. Jones (editors), *Importance, Preservation and Management of Riparian Habitat: A Symposium.* U. S. Department of Agriculture, Forest Service General Technical Report RM-43.

**Brown, D. E. et al. 1980.** *A Digitized Systematic Classification for Ecosystems with an Illustrated Summary of the Natural Vegetation of North America.* U. S. Department of Agriculture, Forest Service General Technical Report RM-73.

**Brownell, H. and Eaton, S. D. 1975.** "The Colorado River Salinity Problem with Mexico," *American Journal of International Law* 69:255-271.

**Bruk, S. and Milorodov, V. 1971.** "Bed Deformation Due to Silting of Nonuniform Sediments in Backwater Affected Rivers," *Transactions of International Association of Hydraulic Research, XV Congress,* Paris.

**Brune, G. H. 1953.** "Troy Efficiency of Reservoirs," *American Geophysical Union Transactions* 34:101-109.

**Brush, L. M. and Wolman, M. G. 1957.** "Knickpoint Behavior in Noncohesive Materials: A Laboratory Study," *Geological Society of America Bulletin* 89: 1489-1498.

**Bryan, K. 1922.** "Date of Channel Trenching Arroyo Cutting in the Arid Southwest," *Science* 62:338-344.

**Bryan, K. 1941.** "Pre-Columbian Agriculture in the Southwest as Conditioned by Periods of Alluviation," *Annals, Association of American Geographers* 31:219-242.

**Bryson R. A. et al. 1970.** "The Character of Late-Glacial and Post-Glacial Climatic Changes," pp. 53-74 in W. Dort, Jr. and J. K. Jones, Jr. (editors), *Pleistocene and Recent Environments of the Central Great Plains.* Lawrence, KS: University of Kansas Press.

**Bull, W. B. 1977.** "The Alluvial Fan Environment," *Progress in Physical Geography* 1:220-270.

**Bull, W. B. and Scott, K. M. 1974.** "Impact of Mining Gravel from Urban Stream Beds in the southern United States," *Geology* 2:171-174.

**Burgess, H. S. and Quirk, J. P. 1978.** *The Theory of the Dam: An Application to the Colorado River.* California Institute of Technology, Division of the Humanities and Social Science, Social Science Working Paper 227.

**Burton, I. et al. 1978.** *The Environment as Hazard.* New York: Oxford University Press.

**Busby, M. W. 1963.** *Yearly Variations in Runoff for the Conterminous United States.* U. S. Geological Survey Water-Supply Paper 1669-S.

**Cabrera, L. 1975.** "Use of the Waters of the Colorado River in Mexico: Pertinent Technical Commentaries," *Natural Resources Journal* 15:27-34.

**Cadigan, R. A. 1970.** *Mercury in the Sedimentary Rocks of the Colorado Plateau Region.* U. S. Geological Survey Professional Paper 713:17-18.

**Casey, H. E. 1972.** *Salinity Problems in Arid Lands Irrigation.* Tuscon, AZ: University of Arizona, Office of Arid Lands Studies.

**Chorley, R. J. and Kennedy, B. A. 1971.** *Physical Geography: A Systems Approach.* London: Prentice-Hall.

**Christensen, E. M. 1962.** "The Rate of Naturalization of *Tamarix* in Utah," *American Midland Naturalist* 68:51-57.

**Church, M. and Mark, D. D. 1980.** "On Size and Scale in Geomorphology," *Progress in Physical Geography* 4:342-390.

**Churchill, M. A. 1948.** "Discussion of Analysis and Use of Reservoir Sedimentation Data' by L. C. Gottschalk," *Proceedings of the Federal Interagency Sedimentation Conference, Denver, Colorado, 1947,* pp. 139-140.

**Clark, A. H. 1956.** "The Impact of Exotic Invasions on the Remaining New World Mid-latitude Grasslands," pp. 737-762 in W. L. Thomas (editor), *Man's Role in Changing the Face of the Earth.* Chicago: University of Chicago Press.

**Clawson, M. and Held, B. 1957.** *The Federal Lands: Their Use and Management.* Lincoln, NE: University of Nebraska Press.

**Clements, W. E. et al. 1978.** *Uranium Mill Tailings as Sources of Atmospheric Rn-222.* Los Alamos Scientific Laboratory Report LA-UR 78-828, Los Alamos, NM.

**Clover, E. A. and Jotter, L. 1944.** "Floristic Studies in the Canyons of the Colorado and Tributaries," *American Midland Naturalist* 32:591-642.

**Cohan, D. R. et al. 1978.** "Avian Population Responses to Salt Cedar Along the Lower Colorado River," pp. 371-381 in R. R. Johnson and J. F. McCormick (editors), *Strategies for Protection and Management of Floodplain Wetlands and Other Riparian Ecosystems.* U. S. Department of Agriculture, Forest Service Research Report WO-12.

**Colton, H. S. 1976.** *The Basaltic Cinder Cones and Lava Flows of the San Francisco Mountain Volcanic Field.* Flagstaff, AZ: Museum of Northern Arizona.

**Condit, W. et al. 1978.** *Sedimentation in Lake Powell.* National Science Foundation, Lake Powell Research Project Report 64, Los Angeles: University of California, Los Angeles.

**Coney, P. J. 1983.** "The Plate Tectonic Setting of Cordilleran Deserts," pp. 81-100 in S. G. Wells and D. R. Haragan (editors), *Origin and Evolution of Deserts.* Albuquerque: University of New Mexico Press.

**Cooke, R. U. et al. 1982.** *Urban Geomorphology in Drylands.* Oxford: Oxford University Press.

**Cooke, R. U. and Reeves, R. W. 1976.** *Arroyos and Environmental Change in the American South-West.* Oxford: Clarendon Press.

**Corle, E. 1951.** *The Gila: River of the Southwest.* Lincoln, NE: University of Nebraska Press.

**Costa, J. E. 1978.** "The Dilemma of Flood Control in the United States," *Environmental Management* 2:313-322.

**Covington, H. R. and Williams, P. L. 1972.** "Map Showing Normal Annual and Monthly Precipitation in the Salina Quadrangle, Utah," U. S. Geological Survey Special Investigations Map I-591-D, Salina Folio.

**Culler, R. C. 1970.** "Water Conservation by Removal of Phreatophytes," *American Geophysical Union Transactions* 51:684-689.

**Culler, R. C. et al. 1982.** *Evapotranspiration Before and After Clearing Phreatophytes, Gila River Flood Plain, Graham County, Arizona.* U. S. Geological Survey Professional Paper 665-P.

**Davis, W. M. 1938.** "Sheetfloods and streamfloods," *Bulletin of the Geological Society of America* 49:329-339.

**Denevan, W. M. 1967.** "Livestock Numbers in Nineteenth-Century New Mexico, and the Problem of Gullying in the Southwest," *Annals,* Association of American Geographers 57:691-703.

**Dingman, S. L. and Platt, R. H. 1977.** "Flood Plain Zoning: Implications of Hydrologic and Legal Uncertainty," *Water Resources Research* 13:519-523.

**Dodge, R. E. 1902.** "Arroyo Formation," *Science* 15:746.

**Dolan, R. et al. 1974.** "Man's Impact on the Colorado River in the Grand Canyon," *American Scientist* 62:392-406.

**Dolan, R. et al. 1978.** "Structural Control of the Rapids and Pools of the Colorado River in the Grand Canyon," *Science* 202:629-631.

**Dunne, T. and Leopold, L. B. 1977.** *Water in Environmental Planning.* San Francisco, CA: W. H. Freeman and Company.

**Dutton, C.E. 1882.** *Tertiary History of the Grand Canyon District.* U. S. Geological Survey Monograph 2.

**Erie, L. J. et al. 1982.** *Consumptive Use of Water by Major Crops in the Southwestern United States.* U. S. Department of Agriculture, Agriculture Research Service, Conservation Research Report 29.

**Evans, N. A. 1975.** "Salt Problem in the Colorado River," *Natural Resources Journal* 15:55-62.

**Faulk, O. B. 1970.** *Arizona: A Short History.* Norman, OK: University of Oklahoma Press.

**Fitzsimmons, J. D. et al. 1980.** "Recent U. S. Population Redistribution: A Geographical Framework for Change in the 1980s," *Social Science Quarterly* 61:485-507.

**Flader, S. L. 1974.** *Thinking Like a Mountain: Aldo Leopold and the Evolution of an Ecological Attitude Toward Deer, Wolves and Forests.* Lincoln, NE: University of Nebraska Press.

**Fortescue, J. A. C. 1980.** *Environmental Geochemistry: A Holistic Approach.* New York: Springer-Verlag.

**Fradkin, P. L. 1981.** *A River No More: The Colorado River and the West.* New York: Alfred A. Knopf.

**Francis J. G. and Ganzel, R. 1984.** *Western Public Lands: The Management of Natural Resources in a Time of Declining Federalism.* Totawa, NJ: Rowman and Allanheld Publishers.

**Fraser, J. C. 1972.** "Regulated Discharge and the Stream Environment," pp. 263-285 in R. T. Oglesby *et al.* (editors), *River Ecology and Man.* New York: Academic Press.

**Freeman, L. R. 1923.** *The Colorado River Yesterday, Today and Tomorrow.* New York: Dodd, Mead and Company.

**Furnish, D. B. and Ladman, J. R. 1975.** "The Colorado River Salinity Agreement of 1973 and the Mexicali Valley," *Natural Resources Journal* 15:83-108.

**Garde, R. J. and Swamee, P. K. 1973.** "Analysis of Aggradation Upstream of a Dam," *International Symposium on River Mechanics.* Bangkok: Thailand.

**Gardner, B. D. and Stewart, C. E. 1978.** "Agriculture and Salinity," pp. 120-131 in D. F. Peterson and A. B. Crawford (editors), *Values and Choices in the Development of the Colorado River Basin.* AZ: University of Arizona Press.

**Gardner, J. S. 1977.** *Physical Geography.* New York: Harper & Row.

**Gatewood, J. S. et al. 1950.** *Use of Water by Bottom-land Vegetation in Lower Safford Valley, Arizona.* U. S. Geological Survey Water-Supply Paper 1103.

**Gay, L. W. and Fritschen, L. J. 1979.** "An Energy Budget Analysis of Water by Saltcedar," *Water Resources Research* 15:1589-1598.

**Gay, L. W. and Hartman, R. K. 1981.** "Energy Budget Measurements Over Irrigated Alfalfa," *Arizona-Nevada Academy of Sciences Proceedings* 11:73-79.

**Gay, L. W. and Hartman, R. K. 1982.** "ET Measurements Over Riparian Saltcedar on the Colorado River," *Hydrology and Water Resources of Arizona and the Southwest* 12:9-15.

**Gessler, J. 1971.** "Aggradation and Degradation," pp. 467-496 in H. W. Shen (editor), *River Mechanics.* Fort Collins, CO: Colorado State University.

**Glancy, P. A. and Harmsen, L. 1975.** *A Hydrologic Assessment of the September 14, 1974, Flood in Eldorado Canyon, Nevada.* U. S. Geological Survey Professional Paper 930.

**Goldwater, B. M. 1970.** *Delightful Journey Down the Green and Colorado Rivers.* Tempe, AZ: Arizona Historical Foundation.

**Graf, W. L. 1978.** "Fluvial Adjustments to the Spread of Tamarisk in the Colorado Plateau Region," *Geological Society of America Bulletin* 89:1491-1501.

**Graf, W. L. 1979a.** "Managing Arizona's Riparian Environments," pp. 21-27 in M. G. Marcus (editor), *Proceedings of the 1979 Summer Conference, Governor's Commission on Arizona Environment.* Phoenix: Governor's Commission on Arizona Environment.

**Graf, W. L. 1979b.** "Rapids in Canyon Rivers," *Journal of Geology* 87:533-551.

**Graf, W. L. 1980a.** "Riparian Management: A Flood Control Perspective," *Journal of Soil and Water Conservation* 35:158-161.

**Graf, W. L. 1980b.** "The Effect of Dam Closure on Downstream Rapids," *Water Resources Research* 16:129-136.

**Graf, W. L. 1982a.** "Tamarisk and River-Channel Management," *Environmental Management* 6:283-296.

**Graf, W. L. 1982b.** "Distance Decay and Arroyo Development in the Henry Mountains Region, Utah," *American Journal of Science* 282:1541-1554.

**Graf, W. L. 1983a.** "Flood-related channel change in an arid-region river," *Earth Surface Processes and Landforms* 8:125-139.

**Graf, W. L. 1983b.** "Downstream Changes in Stream Power in the Henry Mountains, Utah," *Annals,* Association of American Geographers 73:373-387.

**Graf, W. L. 1983c.** "The Arroyo Problem — Paleohydrology and Paleohydraulics in the Short Term," pp. 262-303 in K. J. Gregory (editor), *Background to Paleohydrology.* London: John Wiley and Sons.

**Graf, W. L. 1984.** "A Probabilistic Approach to the Spatial Assessment of River-Channel Instability," *Water Resources Research* 20:953-962.

**Graf, W. L. et al. 1984.** *Issues Concerning Phreatophyte Clearing, Revegetation, and Water Savings Along the Gila River, Arizona.* U. S. Department of the Army, Corps of Engineers, Los Angeles District Contract Report DACW09-83-M-2623.

**Granger, B. H. 1960.** *Will C. Barnes' Arizona Place Names.* Tucson, AZ: University of Arizona Press.

**Gray, E. 1982.** *The Great Uranium Cartel.* Toronto: McClelland and Stewart.

**Gregory, H. E. 1917.** *Geology of the Navajo Country.* U. S. Geological Survey Professional Paper 93.

**Gregory, H. E. 1938.** *The San Juan Country.* U. S. Geological Survey Professional Paper 188.

**Gregory, H. E. 1950.** *Geology and Geography of the Zion Park Region, Utah and Arizona.* U. S. Geological Survey Professional Paper 220.

**Gregory, H. E. and Moore, R. C. 1931.** *The Kaiparowits Region: A Geographic and Geologic Reconnaissance of Parts of Utah and Arizona.* U. S. Geological Survey Professional Paper 164.

**Guy, H. P. and Norman, V. W. 1982.** (reprint) *Field Methods for Measurement of Fluvial Sediment.* Book 3, Chapter C2, Techniques of Water Resources Investigations of the U. S. Geological Survey. Washington, DC: U. S. Government Printing Office.

**Hack, J. T. 1939.** "Late Quaternary History of Several Valleys of Northern Arizona, A Preliminary Announcement," *Museum of Northern Arizona, Museum Notes* 11:63-73.

**Hadley, R. F. 1961.** "Influence of Riparian Vegetation on Channel Shape, Northeastern Arizona," U.S. Geological Survey Professional Paper 424-C, 30-31.

**Hales, J. E. 1974.** "Southwestern United States Summer Monsoon Source — Gulf of Mexico or Pacific Ocean?" *Journal of Applied Meteorology* 13:331-342.

**Hall, S. A. 1977.** "Late Quaternary Sedimentation and Paleoecologic History of Chaco Canyon, New Mexico," *Geological Society of America Bulletin* 88:1593-1618.

**Hammad, H. Y. 1972.** "River Bed Degradation After Closure of Dams," *Proceedings of the American Society of Civil Engineers, Journal of the Hydraulics Division* 98 (HY4):591-607.

**Happ, S. C. 1971.** "Genetic Classification of Valley Sedimentary Deposits," *American Society of Civil Engineers, Hydraulics Division* 97:43-53.

**Harman, J. R. 1971.** *Tropospheric Waves, Jet Streams, and United States Weather Patterns.* Washington, DC: Association of American Geographers, Resource Paper 11.

**Harnack, R. P. and Broccoli, R. J. 1979.** "Associations Between Sea Surface Temperature Gradient and Overlying Mid-Tropospheric Circulation in the North Pacific Region," *Journal of Physical Oceanography* 9:1232-1242.

**Harris, D. R. 1966.** "Recent Plant Invasions in the Arid and Semi-arid Southwest of the United States," *Annals,* Association of American Geographers 56:408-422.

**Harrison, A. S. 1950.** *Report on Special Investigation of Bed Sediment Segregation in a Degrading Bed.* University of California, Berkeley, Institute of Engineering Research, Series 33, Issue 1.

**Hastings, J. R. and Turner, R. M. 1965.** *The Changing Mile.* Tucson, AZ: University of Arizona Press.

**Haury, E. W. 1976.** *The Hohokam.* Tucson, AZ: University of Arizona Press.

**Heede, B. H. 1974.** "Stages of Development of Gullies in Western United States of America," *Zeitschrift fur Geomorphologie (Supplementband)* 18:260-271.

**Heede, B. H. 1976.** *Gully Development and Control: The Status of Our Knowledge.* U.S. Department of Agriculture, Forest Service Research Paper RM-169.

**Hendee, J. C. et al. 1978.** *Wilderness Management.* Washington, DC: U.S. Forest Service.

**Hereford, R. 1984.** "Climate and Ephemeral-Stream Processes: Twentieth Century Geomorphology and Alluvial Stratigraphy of the Little Colorado River, Arizona," *Geological Society of America Bulletin* 95:654-668.

**Herron, W. H. 1917.** *Profile Surveys in the Colorado River Basin in Wyoming, Utah, Colorado, and New Mexico.* U.S. Geological Survey Water-Supply Paper 396.

**Hiibner, C. W. 1974.** "Urbanization in the Colorado River Basin," pp. 216-240 in A. B. Crawford and D. F. Peterson (editors), *Environmental Management and the Colorado River Basin.* Logan, UT: Utah State University.

**Holburt, M. B. and Valentine, V. E. 1972.** "Present and Future Salinity of the Colorado River," *Journal of the Hydraulics Division, Proceedings of the American Society of Civil Engineers* 98 (HY 3):503-520.

**Horton, J. S. 1962.** "Taxonomic notes on *Tamarix Pentandra* in Arizona," *The Southwestern Naturalist* 7:22-28.

**Horton, J. S. 1964.** *Notes on the Introduction of Deciduous Tamarisk.* U.S. Department of Agriculture, Forest Service Research Note RM-16.

**Horton, J. S. 1972.** "Management Problems in Phreatophyte and Riparian Zones," *Journal of Soil and Water Conservation* 27:57-61.

**Horton, J. S. 1977.** "The Development and Perpetuation of the Permanent Tamarisk type in the Phreatophyte Zone of the Southwest," pp. 124-127 in R. R. Johnson and D. A. Jones (editors), *Importance, Preservation, and Management of Riparian Habitat: A Symposium.* U.S. Department of Agriculture, Forest Service General Technical Report RM-43.

**Horton, J. S. and Campbell, C. J. 1974.** *Management of Phreatophytes and Riparian Vegetation for Multiple Use Values.* U.S Department of Agriculture, Forest Service Research Paper RM-117.

**Horton, J. S. et al. 1960.** *Seed Germination and Seedling Establishment of Phreatophyte Species.* U.S. Department of Agriculture, Forest Service Research Paper RM-48.

**Howard, A. and Dolan, R. 1981.** "Geomorphology of the Colorado River in the Grand Canyon," *Journal of Geology* 89:269-298.

**Howard, C. S. 1947.** *Suspended Sediment in the Colorado River.* U.S. Geological Survey Water-Supply Paper 998.

**Howe, E. F. and Hall, W. J. 1910.** *The Story of the First Decade in Imperial Valley.* Imperial, CA: Howe and Sons.

**Huff, L. C. 1970.** *A Geochemical Study of Alluvium-Covered Copper Deposits in Pima County, Arizona.* U.S. Geological Survey Bulletin 1312-C.

**Hughes, W. 1971.** "Effects on Water Supply Due to Salt Cedar Removal," American Society of Civil Engineers, National Water Resources Engineering Meeting, Phoenix, Preprint 1290.

**Hundley, N. 1964.** "The Colorado River Dispute," *Foreign Affairs* 42:495-500.

**Hundley, N. 1966.** *Dividing the Waters.* Berkeley, CA: University of California Press.

**Hunt, C. B. 1972.** *Geology of Soils — Their Evolution, Classification, and Uses.* San Francisco, CA: W. H. Freeman and Company.

**Hunt, C. B. 1974.** *Natural Regions of the United States and Canada.* San Francisco, CA: W. H. Freeman and Company.

**Hunt, C. B. et al. 1953.** *Geology and Geography of the Henry Mountains Region, Utah.* U.S. Geological Survey Professional Paper 228.

**Huntington, E. 1914.** *The Climatic Factor as illustrated in Arid America.* Washington, DC: Carnegie Institution of Washington.

**International Commission on Large Dams. 1973.** *World Register of Dams.* Paris, France: International Commission on Large Dams.

**Iorns, W. V. et al. 1964.** *Water Resources of the Upper Colorado River Basin — Basic Data.* U.S. Geological Survey Professional Paper 442.

**Iorns, W. V. et al. 1965.** *Water Resources of the Upper Colorado River Basin — Technical Report.* U.S. Geological Survey Professional Paper 441.

**Jacoby, G. C. et al. 1976.** "Law, Hydrology, and Surface Water Supply in the Upper Colorado River Basin," *American Resources Bulletin* 16:102-121.

**Johnson, R. 1977.** *The Central Arizona Project, 1918-1968.* Tucson, AZ: University of Arizona Press.

**Johnson, R. R. 1970.** "Tree Removal Along Southwestern Rivers and Effects on Associated Organisms," *American Philosophical Society 1980 Yearbook,* 321-322.

**Kates, R. W. 1962.** *Hazard and Choice Perception in Flood Plain Management.* Chicago: University of Chicago Press.

**Kaufman, R. F. et al. 1976.** "Effects of Uranium Mining and Milling on Ground Water in the Grants Mineral Belt, New Mexico," *Ground Water* 14:296-308.

**Keller, E. A. 1976.** "Channelization: environmental, geomorphic, and engineering apsccts," pp. 115-140 in D. R. Coates (editor), *Geomorphology and Engineering.* Stroudsburg, PA: Dowden, Hutchingson, and Ross.

**Knox, J. C. 1978.** "Arroyos and Environmental Change in the American South-West," Book Review, *Annals,* Association of American Geographers 68:137-139.

**Knox, J. C. 1983.** "Responses of River Systems to Holocene Climates," pp. 26-41 in H. E. Wright (editor), *Late-Quaternary Environments of the United States, Volume 2 — The Holocene.* Minneapolis: University of Minnesota Press.

**Kunkler, J. L. 1979.** *A Reconnaissance Study of Selected Environmental Impacts on Water Resources Due to the Exploration, Mining, and Milling of Uraniferous Ores in the Grants Mineral Belt, Northwestern New Mexico.* Albuquerque, NM: San Juan Basin Regional Uranium Study, U.S. Geological Survey.

**Lane, E. W. 1957.** *A Study of the Shape of Channels Formed by Natural Streams Flowing in Erodible Material.* Missouri River Division Sediment Series No. 9, Omaha, NE: U.S. Army Corps of Engineers.

**LaRue, E. C. 1916.** *The Colorado River and Its Utilization.* U.S. Geological Survey Water-Supply Paper 395.

**LaRue, E. C. 1925.** *Water Power and Flood Control of the Colorado River Below Green River, Utah.* U.S. Geological Survey Water-Supply Paper 556.

**Larson, D. K. 1974.** "An Analysis of the Motor-Row Conversion Issue of Colorado River Float Trips," Unpublished Thesis, University of Arizona, Tucson, AZ.

**Lattman, L. H. 1960.** "Cross-section of a floodplain in a moist region of moderate relief," *Journal of Sedimentary Petrology* 30:275-282.

**Laursen, E. M. et al. 1976.** "On Sediment Transport Through the Grand Canyon," pp. 4/76-4/87 in *Proceedings of the Third Interagency Sedimentation Conference.* Denver, CO.

**Leopold, A. 1921.** "A Plea for Recognition of Artificial Works in Forest Erosion and Control Policy," *Journal of Forestry* 19:267-273.

**Leopold, L. B. 1951.** "Rainfall Frequency: An Aspect of Climatic Variation," *Transactions of the American Geophysical Union* 32:347-357.

**Leopold, L. B. 1969.** *The Rapids and Pools — Grand Canyon.* U.S. Geological Survey Professional Paper 669-D.

**Leopold, L. B. and Bull, W. B. 1979.** "Base Level, Aggradation, and Grade," *Proceedings of the American Philosophical Society* 123:168-202.

**Leopold, L. B. and Wolman, M. G. 1957.** *River Channel Patterns: Braided, Meandering, and Straight.* U.S. Geological Survey Professional Paper 282-B.

**Leopold, L. B. et al. 1964.** *Fluvial Processes in Geomorpholoy.* San Francisco, CA: W. H. Freeman.

**Levinson, A. A. 1980.** *Introduction to Exploration Geochemistry.* Wimette, IL: Applied Publishing.

**Lewin, J. 1978.** "Floodplain geomorphology," *Progress in Physical Geography* 2:408-437.

**Lewin, J. et al. 1977.** "Interactions Between Channel Change and Historic Mining Sediments," pp. 353-368 in K.J. Gregory (editor), *River Channel Change.* New York: John Wiley and Sons.

**Lingenfelter, R. E. 1978.** *Steamboats on the Colorado River.* Tucson, AZ: University of Arizona Press.

**Little, W. C. and Mayer, P. G. 1972.** *The Role of Sediment Gradation on Channel Armoring.* Georgia Institute of Technology, School of Civil Engineering Report.

**Love, D. W. 1979.** "Quaternary Fluvial Geomorphic Adjustments in Chaco Canyon, New Mexico," pp. 277-308 in D. D. Rhodes and G. P. Williams (editors), *Adjustments of the Fluvial System.* Dubuque, IA: Kendall/Hunt.

**Lusby, G. C. et al. 1971.** *Effects of Grazing on the Hydrology and Biology of the Badger Wash Basin in Western Colorado, 1953-66.* U.S. Geological Survey Water-Supply Paper 1532-D.

**Marcus, W. A. 1983.** "Copper Dispersion in Ephemeral Stream Sediments, Queen Creek, Arizona," unpublished M.A. Thesis, Arizona State University, Department of Geography.

**Martin, P. S. 1963.** *The Last 10,000 Years: A Fossil Pollen Record of the American Southwest.* Tucson, AZ: University of Arizona Press.

**Martin, W. E. et al. 1984.** *Saving Water in a Desert City.* Baltimore, MD: Johns Hopkins University for Resources for the Future.

**McClintock, E. 1951.** "Studies in California Ornamental Plants, 3. The Tamarisks," *Journal of the California Horticultural Society* 12:76-83.

**McGee, W J 1897.** "Sheetflood Erosion," *Bulletin of the Geological Society of America* 8:87-112.

**Meredith, H. L. 1968.** "Reclamation in the Salt River Valley, 1902-1917," *Journal of the West* 7:76-83.

**Mermel, T. W. (editor) 1958.** *Register of Dams in the United States; Completed, Under Construction, and Proposed.* New York: McGraw-Hill Book Company.

**Merriam, C. H. 1890.** *Results of a Biological Survey of the San Francisco Mountain Region and Desert of the Little Colorado, Arizona.* Washington, DC: Bureau of the Biological Survey.

**Merritt, R. C. and Pings, W. B. 1969.** *Processing of Uranium Ores.* Mineral Industries Bulletin 12 of the Colorado School of Mines Research Institute, Golden, CO.

**Merritt, R. C. and Pings, W. B. 1971.** *The Extractive Metallurgy of Uranium.* Boulder, CO: Johnston Publishing Company.

**Miall, A. D. 1977.** "A Review of the Braided River Depositional Environment," *Earth Science Reviews* 13:1-62.

**Miller, C. R. et al. 1962.** "Upland Gully Sediment Production,'" International Association of Scientific Hydrology, Commission on Land Erosion, Publication 59, 83-104.

**Miller, D. E. 1966.** *Hole-in-the-Rock: An Epic in the Colonization of the Great American West.* Salt Lake City, UT: University of Utah Press.

**Miller, H. 1970.** "Methods and Results of River Terracing," pp. 45-63 in G. H. Dury (editor), *Rivers and River Terraces.* London: Macmillan.

**Miser, H. D. 1924a.** *Geological Structure of San Juan Canyon and Adjacent Country, Utah.* U.S. Geological Survey Bulletin 751-B.

**Miser, H. D. 1924b.** *The San Juan Canyon, Southeastern Utah: a Geographic and Hydrographic Reconnaissance.* U.S. Geological Survey Water-Supply Paper 538.

**Mitchell, V. L. 1976.** "The Regionalization of Climate in the Western United States," *Journal of Applied Meteorology* 15:920-927.

**Murray, J. A. et al. 1933.** *The Oxford English Dictionary, Being a Corrected Re-Issue With an Introduction, Supplement, and Bibliography of a New English Dictionary on Historical Principles.* Oxford: Clarendon Press.

**Nadeau, R. A. 1974.** *The Water Seekers.* Salt Lake City, UT: Peregrine Smith.

**Namias, J. 1969.** "Seasonal Interactions Between the North Pacific Ocean and the Atmosphere During the 1960's," *Monthly Weather Review* 97:173-192.

**Nash, R. 1967.** *Wilderness and the American Mind.* New Haven, CT: Yale University.

**Nash, R. 1970.** *Grand Canyon of the Living Colorado.* New York: Sierra Club and Ballantine Books.

**Neff, E. L. 1967.** "Discharge Frequency Compared to Long Term Sediment Yields," *Publications of the International Association of Scientific Hydrology* 75:236-242.

**Nelson, J. D. and Shepherd, T. A. (editors). 1982.** *Uranium Mill Tailings Management.* Fort Collins, CO: Colorado State University.

**Office of Emergency Preparedness, 1972.** *Disaster Preparedness.* Washington, DC: U.S. Government Printing Office.

**Oka, J. N. 1962.** *Sedimentation Study, Glen Canyon Dam, Colorado River Storage Project.* Salt Lake City, UT: U.S. Bureau of Reclamation.

**Park, C. C. 1981.** "Man, River Systems, and Environmental Impact," *Progress in Physical Geography* 5:1-31.

**Pase, C. P. and Layser, E. F. 1977.** "Classification of Riparian Habitat in the Southwest," pp. 5-9 in R. R. Johnson and D. A. Jones (editors), *Importance, Preservation and Management of Riparian Habitat: A Symposium.* U.S. Department of Agriculture, Forest Service General Technical Report RM-43.

**Patric, W. C. 1981.** *Trust Land Administration in the Western Lands.* Denver, CO: Public Lands Institute.

**Patten, D. T. 1984.** "Revegetation Evaluation," pp. 25-50 in W. L. Graf, *et al.* (editors), *Issues Concerning Phreatophyte Clearing, Revegetation, and Water Savings Along the Gila River, Arizona.* U.S. Army Corps of Engineers, Los Angeles District Contract Report DACW09-83-M-2623, 25-50.

Pearthree, M. S. 1982. "Channel Change in the Rillito Creek System; Southeastern Arizona: Implications for Floodplain Management," unpublished MS thesis, University of Arizona, Tucson, AZ.

Pemberton, E. L. 1976. "Channel Changes in the Colorado River Below Glen Canyon Dam," pp. 5/61-5/73 in *Proceedings of the Third Interagency Sedimentation Conference, Denver, Colorado.*

Perkins, B. L. 1979. *An Overview of the New Mexico Uranium Industry.* Santa Fe, NM: New Mexico Energy and Minerals Department.

Pickard, J. P. 1973. "Growth of Urbanized Population in the United States: Past, Present, and Future," pp. 10-21 in D. W. Rasmussen and C. T. Haworth (editors), *The Modern City: Readings in Urban Economics.* San Francisco, CA: Harper and Row.

Pickup, G. *et al.* 1983. "Modelling Sediment Transport as a Moving Wave — The Transfer and Deposition of Mining Waste," *Journal of Hydrology* 60:281-301.

Porter, E. 1963. *The Place No One Knew.* San Francisco, CA: Sierra Club Books.

Porterfield, G. 1972. *Computation of Fluvial Sediment Discharge.* Book 3, Chapter C3, *Techniques of Water Resources Investigations of the U.S. Geological Survey.* Washington, DC: U.S. Government Printing Office.

Potter, L. D. *et al.* 1975. "Mercury Levels in Lake Powell," Environmental Science and Technology 9:41-46.

Powell, J. W. 1893. *Thirteenth Annual Report of the U.S. Geological Survey.* Washington, DC: Department of Interior.

Rahn, P. H. 1967. "Sheetfloods, Streamfloods, and the Formation of Pediments," *Annals,* Association of American Geographers 57:593-604.

Rhoads, B. L. 1984. "Flood Hazard Assessment for Land Use Planning Near Desert Mountains," *Environmental Management,* in press.

Rhodes, S. L. *et al.* 1984. "Climate and the Colorado River: the Limits of Management," *Bulletin of the American Meteorological Society* 65:682-691.

Rich, J. L. 1911. "Recent Stream Trenching in the Semi-Arid Portion of Southwestern New Mexico, a Result of Removal of Vegetation Cover," *American Journal of Science* 32:237-245.

Ritter, D. F. 1978. *Process Geomorphology.* Dubuqe, IA: W. C. Brown.

Robbins, R. M. 1942. *Our Landed Heritage: The Public Domain, 1776-1936.* Princeton, NJ: Princeton University Press.

Robinson, M. C. 1979. *Water for the West: The Bureau of Reclamation, 1902-1977.* Chicago: Public Works Historical Society.

Robinson, T. W. 1958. *Phreatophytes.* U. S. Geological Survey Water-Supply Paper 1423, 70-75.

Robinson, T. W. 1965. *Introduction, Spread, and Areal Extent of Saltcedar (*Tamarix*) in the Western States.* U.S. Geological Survey Professional Paper 491-A.

Rohrbach, M. J. 1968. *The Land Office Business: The Settlement and Administration of American Public Lands, 1789-1837.* New York: Oxford University Press.

Rutter, E. J. and Engstrom, L. R. 1964. "Hydrology of Flow Control: Reservoir Regulation," pp. 25/60-25/97 in V.T. Chow (editor), *Handbook of Hydrology.* New York: McGraw-Hill Book Company.

St. Anthony Falls Laboratory. 1957. *A Study of Methods Used in Measurement and Analysis of Sediment Loads in Streams.* Minneapolis: St. Anthony Falls Laboratory.

Schumm, S. A. 1973. "Geomorphic Thresholds and Complex Response of Drainage Systems," pp. 299-310 in M. Morisawa (editor), *Fluvial Geomorphology.* Binghamton, NY: State University of New York at Binghamton.

Schumm, S. A. 1976. "Episodic Erosion, A Modification of the Geomorphic Cycle," pp. 69-86 in W. N. Melhorn and R. C. Flemal (editors), *Theories of Landform Development.* Binghamton, NY: State University of New York at Binghamton.

**Schumm, S. A. 1977.** *The Fluvial System.* New York: John Wiley and Sons.
**Schumm, S. A. and Hadley, R. F. 1957.** "Arroyos and the Semi-Arid Cycle of Erosion," *American Journal of Science* 255:161-174.
**Schumm, S. A. et al. 1972.** "Variability of River Patterns," *Nature, Physical Science* 237:75-76.
**Schumm, S. A. and Lichty, R. W. 1963.** *Channel Widening and Flood-Plain Construction Along Cimarron River in Southwestern Kansas.* U.S. Geological Survey Professional Paper 352-D.
**Sebenik, P. G. and Thames, J. L. 1967.** "Water Consumption by Phreatophytes," *Progressive Agriculture in Arizona* 19:10-11.
**Sellers, W. D. and Hill, R. H. 1974.** *Arizona Climate 1931-1972.* Tucson, AZ: University of Arizona Press.
**Sheaffer, J. R. 1967.** *Introduction to Flood Proofing.* Chicago: University of Chicago Press.
**Sheridan, D. 1981.** *Desertification of the United States.* Washington, DC: Council on Environmental Quality.
**Shulits, S. 1941.** "Rational Equation of River-Bed Profile," *American Geophysical Union Transactions* 22:622-630.
**Simons, D. B. and Senturk, F. 1976.** Sediment Transport Technology. Fort Collins, CO: Water Resources Publications.
**Sittig, M. (editor). 1981.** *Priority Toxic Pollutants: Health Impacts and Allowable Limits.* Park Ridge, NJ: Noyes Data Corporation.
**Smith, C. L. 1972.** *The Salt River Project: A Case Study in Cultural Adaptation to an Urbanizing Community.* Tucson, AZ: University of Arizona Press.
**Smith, D. E. 1981.** "Riparian Vegetation and Sedimentation in a Braided River," unpublished M.A. thesis, Arizona State University.
**Smith, K. and Tobin, G. A. 1979.** *Topics in Applied Geography: Human Adjustment to the Flood Hazard.* London: Longman Group Limited.
**Smith, W. O. et al. 1960.** *Comprehensive Survey of Sedimentation in Lake Mead, 1948-1949.* U.S. Geological Survey Professional Paper 295.
**Snow, A. 1977.** (Third Edition). *Rainbow Views: A History of Wayne County.* Springville, UT: Art City Publishers.
**Standiford, D. R. et al. 1973** *Mercury in the Lake Powell Ecosystem.* National Science Foundation, Lake Powell Research Project Bulletin 1, Los Angeles: University of California, Los Angeles.
**Steen, H. K. 1976.** *The U.S. Forest Service: A History.* Seattle, WA: University of Washington Press.
**Stegner, W. (editor). 1955.** *This is Dinosaur.* New York: Alfred A. Knopf.
**Stelezer, K. 1981.** *Bed Load Transport: Theory and Practice.* Littletown, CO: Water Resources Publications.
**Stockton, C. W. 1975.** *Long-Term Streamflow Records Reconstructed From Tree-Rings.* Tucson, AZ: University of Arizona Press, Paper 5 of the Laboratory of Tree Ring Research.
**Strahler, A. N. 1956.** "The Nature of Induced Erosion and Aggradation," vol. 2, pp. 621-638 in W. L. Thomas, Jr. (editor), *Man's Role in Changing the Face of the Earth.* Chicago: University of Chicago Press.
**Summerfeld, M. R. et al. 1976.** *Survey of Bacteria, Phytoplankton, and Trace Chemistry of the Lower Colorado River and Tributaries in the Grand Canyon National Park.* U.S. National Park Service, Colorado River Research Program Report 40, Denver, CO.
**Sutton, I. 1968.** "Geographical Aspects of Construction Planning: Hoover Dam Revisited," *Journal of the West* 7:301-344.
**Swift, T. T. 1926.** "Rate of Channel Trenching in the Southwest," *Science* 63:70-71.

**Taylor, R. W. and Taylor, S. W. 1970.** *Uranium Fever, or No Talk Under $1 Million.* New York: Macmillan Company.

**Terrell, J. W. 1965.** *War for the Colorado River: Volume 1, The California-Arizona Controversy.* Glendale, CA: Arthur H. Clark Company.

**Thornbury, W. D. 1965.** *Regional Geomorphology of the United States.* New York: John Wiley and Sons.

**Thornthwaite, C. W. and Mather, J. R. 1955.** *The Water Balance.* Centerton, NJ: Laboratory of Climatology.

**Thornthwaite, C. W. and Mather, J. R. 1957.** *Instructions and Tables for Computing Potential Evapotranspiration and the Water Balance.* Centerton, NJ: Laboratory of Climatology.

**Tomanek, G. W. and Ziegler, R. L. 1961.** *Ecological Studies of Saltcedar.* Fort Hays, KS: Fort Hays State College, Botany Department.

**Towler, S. 1982.** *Cocopele Stories.* Deadwood, OR: Coyote Press.

**Trelease, F. J. 1979.** *Cases and Materials on Water Law (3rd Edition).* St. Paul, MN: West Publishing Company.

**Turner, D. S. 1971.** "Dams and Ecology," *Civil Engineering* 41:76-80.

**Turner, R. M. 1974.** *Quantitative and Historical Evidence of Vegetation Changes Along the Upper Gila River, Arizona.* U.S. Geological Survey Professional Paper 655-H.

**Turner, R. M. and Karpiscak, M. M. 1980.** *Recent Vegetation G Changes Along the Colorado River Between Glen Canyon Dam and Lake Mead, Arizona.* U.S. Geological Survey Professional Paper 1132.

**U.S. Army. 1979a.** *Flood Damage Report, February, 1979.* Los Angeles: Corps of Engineers, Los Angeles District.

**U.S. Army. 1979b.** *Flood Damage Report, Phoenix Metropolitan Area, December, 1978.* Los Angeles: Corps of Engineers, Los Angeles District.

**U.S. Army. 1980.** *Phoenix Flood Damage Survey, February 1980.* Los Angeles: Corps of Engineers, Los Angeles District.

**U.S. Army. 1981.** *HEC-2, Water Surface Profiles, Users Manual.* Davis, CA: Hydraulic Engineering Center.

**U.S. Army Corps of Engineers. 1956.** *Snow Hydrology.* Washington, DC: Office of Technical Services, Department of Commerce.

**U.S. Army Corps of Engineers and U.S. Bureau of Reclamation. 1982.** *Colorado River Basin — Hoover Dam: Review of Flood Control Regulation, Final Report.* Los Angeles: Corps of Engineers.

**U.S. Bureau of Land Management. 1978.** *Public Land Statistics.* Washington, DC: U.S. Department of Interior.

**U.S. Bureau of Reclamation. 1946.** *The Colorado River: A Natural Menace Becomes a National Resource.* Washington, DC: U.S. Department of Interior.

**U.S. Bureau of Reclamation. 1973a.** *Evapotranspirometer Studies of Saltcedar Near Bernardo, New Mexico, March 1973.* Albuquerque, NM: Pacific Southwest Interagency Commission.

**U.S. Bureau of Reclamation. 1973b.** *Progress Report: Phreatophyte Investigations, Bernardo Evapotranspirometers.* Albuquerque, NM: Rio Grande Project Office, Bureau of Reclamation.

**U.S. Bureau of Reclamation. 1975.** *Final Environmental Impact Statement, Colorado Basin Salinity Control Project, Title 1.* Washington, DC: Bureau of Reclamation.

**U.S. Environmental Protection Agency. 1975.** *Water Quality Impacts of Uranium Mining and Milling Activities in the Grants Mineral Belt, New Mexico.* Springfield, VA: Environmental Protection Agency Report 906/9-75-002, National Technical Information Service.

**U.S. Environmental Protection Agency. 1976.** *Quality Criteria for Water.* Washington, DC: Government Printing Office.

**U.S. Fish and Wildlife Service. 1976.** *The Southern Bald Eagle in Arizona: A Status Report.* U.S. Department of Interior, Fish and Wildlife Service, Endangered Species Report 1.

**U.S. Geological Survey. 1970.** *The National Atlas of the United States of America.* Washington, DC: U.S. Department of Interior.

**U.S. National Park Service. 1977a.** *Draft Management Plan and Environmental Impact Assessment, Glen Canyon National Recreation Area, Arizona and Utah.* Denver, CO: National Park Service.

**U.S. National Park Service. 1977b.** *Draft Management Plan and Environmental Impact Assessment, Dinosaur National Monument, Colorado and Utah.* Denver, CO: National Park Service.

**U.S. National Park Service. 1977c.** *Draft Management Plan and Environmental Impact Assessment, Colorado River Unit, Grand Canyon National Park, Arizona.* Denver, CO: National Park Service.

**U.S. National Park Service. 1977d.** *Assessment of Alternatives, General Management Plan, Canyonlands National Park, Utah.* Denver, CO: National Park Service.

**U.S. National Park Service. 1979.** *Final Environmental Statement, Proposed Colorado River Management Plan, Grand Canyon, Arizona.* Denver, CO: National Park Service.

**U.S. Water Resources Council. 1977a.** *'75 Water Assessment, Lower Colorado Region, Summary Report.* Washington, DC: Water Resources Council.

**U.S. Water Resources Council. 1977b.** *'75 Water Assessment, Upper Colorado Region, Summary Report.* Washington, DC: Water Resources Council.

**Van Devender, T. R. and Spaulding, W. G. 1979.** "Development of Vegetation in the Southwestern United States," *Science* 204:701-710.

**van Hylckama, T. E. A. 1974.** *Water Use by Saltcedar as Measured by the Water Budget Method.* U.S. Geological Survey Professional Paper 491-E.

**Ward, J. V. and Stanford, J. A. 1979.** *The Ecology of Regulated Streams.* New York: Plenum Press.

**Ward, R. 1978.** *Floods: A Geographical Perspective.* New York: John Wiley & Sons.

**Warne, W. E. 1973.** *The Bureau of Reclamation.* New York: Praeger Books.

**Warren, D. K. and Turner, R. M. 1975.** "Saltcedar (*Tamarix chinensis*) Seed Production, Seedling Establishment, and Response to Inundation," *Arizona Academy of Science Journal* 10:135-144.

**Waters, F. 1963.** *Book of the Hopi.* New York: Penguin Books.

**Waters, F. 1946.** *The Colorado.* New York: Holt, Rinehart Winston.

**Wedepohl, A. A. 1978.** *Handbook of Geochemistry.* New York: Springer.

**Weimer, W. C. *et al.* 1981.** *Survey of Radionuclide Distributions Resulting from the Church Rock, New Mexico Uranium Mill Tailings Pond Dam Failure.* Nuclear Regulatory Commission Report CR-2449, PNL-4122.

**White, G. F. 1964.** *Choice of Adjustment to Floods.* Chicago: University of Chicago Press.

**Whitewater Committee. 1982.** *River Information Digest For Popular Western Whitewater Boating Rivers Managed by Federal Agencies.* Lewiston, ID: National Park Service, Bureau of Land Management, and Forest Service.

**Wildan Associates. 1981.** *Agua Fria River Study — 1981.* Flood Control District of Maricopa County, Arizona, Contract Report FCD 81-2.

**Wilkening, M. H. *et al.* 1972.** "Radon-222 Flux Measurements in Widely Separated Regions," *Proceedings of the Second International Symposium on the Natural Radiation Environment,* Rice University and University of Texas School of Public Health, Houston, TX, 717-730.

**Williams, G. P. and Wolman, M. G. 1984.** *Downstream Effects of Dams on Alluvial Rivers.* U.S. Geological Survey Professional Paper 1286.

**Witzig, B. J. 1944.** "Sedimentation in Reservoirs," *American Society of Civil Engineers, Transactions* 109: 1047-1071.

**Wolfenden, P. J. and Lewin, J. 1978.** "Distribution of Metal Pollutants in Active Stream Sediments," *Catena* 5:67-78.

**Woolheiser, D. A. and Lenz, A. T. 1965.** "Channel Gradients Above Gully-Control Structures," *American Society of Civil Engineers, Journal of the Hydraulics Division,* Paper 4333, 165-187.

**Woodbury, A. M. 1944.** "A History of Southern Utah and its National Park," *Utah State Historical Society* 3-4:111-223.

**Wyant, W. K. 1982.** *Westward in Eden.* Berkeley, CA: University of California Press.

**Yang, C. T. and Song, C. C. S. 1979.** "Dynamics of Adjustments of Alluvial Channels," pp. 55-68 in D. D. Rhodes and G. P. Williams (editors), *Adjustments of the Fluvial System.* Dubuque, IA: Kendall/Hunt Publishers.